CONNECTIONS

The EERI Oral History Series

Vitelmo V. Bertero

P9-DTE-818

CONNECTIONS
The EERI Oral History Series

Vitelmo V. Bertero

Robert Reitherman, Interviewer

 Earthquake Engineering Research Institute

Editor: Gail Hynes Shea, Berkeley, California, www.gailshea.com

Cover and book design: Laura H. Moger, Moorpark, California, www.lauramoger.com

Copyright 2009 by the Earthquake Engineering Research Institute

The publication of this book was supported by FEMA/U.S. Department of Homeland Security under grant #EMW-2008-CA-0625.

All rights reserved. All literary rights in the manuscript, including the right to publish, are reserved to the Earthquake Engineering Research Institute. No part may be reproduced, quoted, or transmitted in any form without the written permission of the Executive Director of the Earthquake Engineering Research Institute. Requests for permission to quote for publication should include identification of the specific passages to be quoted, anticipated use of the passages, and identification of the user.

The opinions expressed in this publication are those of the oral history subject and do not necessarily reflect the opinions or policies of the Earthquake Engineering Research Institute.

Published by the Earthquake Engineering Research Institute

>   499 14th Street, Suite 320
>   Oakland, CA 94612-1934
>   Tel: (510) 451-0905   Fax: (510) 451-5411
>   E-mail: eeri@eeri.org
>   Website: http://www.eeri.org

EERI Publication Number: OHS-16

Library of Congress Cataloging-in-Publication Data
Bertero, Vitelmo V. (Vitelmo Victorio)
 Vitelmo V. Bertero / Robert Reitherman, interviewer.
    p. cm. -- (Connections : the EERI oral history series)
 Includes index.
 "Sponsored by the Earthquake Engineering Research Institute with support
from FEMA/U.S. Department of Homeland Security."
 ISBN 978-1-932884-42-5 (alk. paper)
 1. Bertero, Vitelmo V. (Vitelmo Victorio) 2. Civil engineers--United
States--Interviews. 3. Earthquake engineering. I. Reitherman, Robert,
1950- II. Title.
 TA140.B425A3 2009
 624.092--dc22
                    2009014148

Printed in the United States of America

1 2 3 4 5 6 7 8 20 19 18 17 16 15 14 13 12 11 10 09

Table of Contents

The EERI Oral History Series

This is the sixteenth volume in the Earthquake Engineering Research Institute's *Connections: The EERI Oral History Series*. EERI began this series to preserve the recollections of some of those who have had pioneering careers in the field of earthquake engineering. Significant, even revolutionary, changes have occurred in earthquake engineering since individuals first began thinking in modern, scientific ways about how to protect construction and society from earthquakes. The *Connections* series helps document this important history.

Connections is a vehicle for transmitting the fascinating accounts of individuals who were present at the beginning of important developments in the field, documenting sometimes little-known facts about this history, and recording their impressions, judgments, and experiences from a personal standpoint. These reminiscences are themselves a vital contribution to our understanding of where our current state of knowledge came from and how the overall goal of reducing earthquake losses has been advanced. The Earthquake Engineering Research Institute, incorporated in 1948 as a nonprofit organization to provide an institutional base for the then-young field of earthquake engineering, is proud to help tell the story of the development of earthquake engineering through the *Connections* series. EERI has grown from a few dozen individuals in a field that lacked any significant research funding to an organization with about 2,500 members today. It is still devoted to its original goal of investigating the effects of destructive earthquakes and publishing the results through its reconnaissance report series. EERI brings researchers and practitioners together to exchange information at annual meetings and, via a now-extensive calendar of conferences and workshops, provides a forum through which individuals and organizations of various disciplinary backgrounds can work together for increased seismic safety.

The EERI oral history program was initiated by Stanley Scott (1921-2002). The first nine volumes were published during his lifetime, and manuscripts and interview transcripts he left to EERI are resulting in the publication of other volumes for which he is being posthumously credited. In addition, the Oral

History Committee is including further interviewees within the program's scope, following the Committee's charge to include subjects who: 1) have made an outstanding career-long contribution to earthquake engineering, 2) have valuable first-person accounts to offer concerning the history of earthquake engineering, and 3) whose backgrounds, considering the series as a whole, appropriately span the various disciplines that are included in the field of earthquake engineering.

Scott's work, which he began in 1984, summed to hundreds of hours of taped interview sessions and thousands of pages of transcripts. Were it not for him, valuable facts and recollections would already have been lost.

Scott was a research political scientist at the Institute of Governmental Studies at the University of California at Berkeley. He was active in developing seismic safety policy for many years, and was a member of the California Seismic Safety Commission from 1975 to 1993. Partly for that work, he received the Alfred E. Alquist Award from the Earthquake Safety Foundation in 1990.

Scott received assistance in formulating his oral history plans from Willa Baum, then Director of the University of California at Berkeley Regional Oral History Office, a division of the Bancroft Library. An unfunded interview project on earthquake engineering and seismic safety was approved, and Scott was encouraged to proceed. Following his retirement from the University in 1989, Scott continued the oral history project. For a time, some expenses were paid from a small grant from the National Science Foundation, but Scott did most of the work pro bono. This work included not only the obvious effort of preparing for and conducting the interviews themselves, but also the more time-consuming tasks of reviewing transcripts and editing the manuscripts to flow smoothly.

The *Connections* oral history series presents a selection of senior individuals in earthquake engineering who were present at the beginning of the modern era of the field. The term "earthquake engineering" as used here has the same meaning as in the name of EERI—the broadly construed set of disciplines, including geosciences and social sciences as well as engineering itself, that together form a related body of knowledge and collection of individuals that revolve around the subject of earthquakes. The events described in the *Connections* series span many kinds of activities: research, design projects, public policy, broad social aspects, and education, as well as interesting personal aspects of the subjects' lives.

Published volumes in Connections: *The EERI Oral History Series*

Henry J. Degenkolb	1994
John A. Blume	1994
Michael V. Pregnoff and John E. Rinne	1996
George W. Housner	1997
William W. Moore	1998
Robert E. Wallace	1999
Nicholas F. Forell	2000
Henry J. Brunnier and Charles De Maria	2001
Egor P. Popov	2001
Clarence R. Allen	2002
Joseph Penzien	2004
Robert Park and Thomas Paulay	2006
Clarkson W. Pinkham	2006
Joseph P. Nicoletti	2006
LeRoy Crandall	2008
Vitelmo V. Bertero	2009

EERI Oral History Committee

Robert Reitherman, Chair
Thalia Anagnos
William Anderson
Roger Borcherdt
Gregg Brandow
Ricardo Dobry
Robert D. Hanson
Loring A. Wyllie, Jr.

Foreword

This oral history volume is the culmination of the interview sessions Vitelmo Victorio Bertero and I had in 2006, 2007, and 2008. Two members of the Oral History Committee, Loring Wyllie and Robert D. Hanson, reviewed the manuscript and improved it with corrections, as did EERI President Thalia Anagnos. The author of the Personal Introduction, Joseph Penzien, also reviewed the manuscript and spotted some needed factual corrections that slipped by the rest of us.

Gail Shea, consulting editor to EERI, carefully reviewed the entire manuscript and prepared the index, as she has on previous *Connections* volumes, and Eloise Gilland, the Editorial and Publications Manager of EERI, also assisted in seeing this publication through to completion.

Robert Reitherman
Chair, EERI Oral History Committee
November 2008

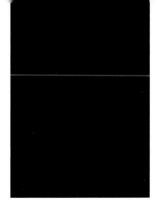

Personal Introduction

It is indeed an honor to contribute this Personal Introduction of Vitelmo V. Bertero to the sixteenth volume of *Connections: The EERI Oral History Series.* He and I have followed similar paths over the past five decades in advancing our separate careers in the general field of earthquake engineering. These similarities include: 1) being born and raised on farms; he in Argentina, I in South Dakota, 2) following the same strong interests in mathematics and science while in high school, 3) attending graduate school in the Department of Civil Engineering at MIT, where we both received ScD degrees, 4) joining the faculty of Civil Engineering at the University of California, Berkeley in the 1950s, 5) teaching courses mainly in structural engineering and conducting research with emphasis on the performance of structures under seismic conditions, and 6) providing specialty consulting services to the engineering profession, focusing primarily on seismic-related problems.

In our doctoral programs at MIT, Professor Bertero and I both conducted experimental research on blast effects on reinforced concrete structures under the guidance of Professor Robert Joseph Hansen. Having been at MIT (1947-1950) somewhat before Professor Bertero, I was assigned the task of focusing on reinforced concrete beams, while he, being at MIT somewhat later (1953-1957), was assigned the more difficult task of focusing on reinforced concrete shear walls. When I first got involved, Professor Hansen had already designed the testing machine for applying a blast-type impulsive load to beams. However, when Professor Bertero got involved, the more complex testing machine for applying such a load to shear walls had not yet been fully developed. Thus, he was assigned a much more difficult task than I had been given. Nevertheless, he completed the testing machine design, perfected its performance, and conducted his research superbly, providing much valuable information for the design of blast-resistant underground shelters. I am sure this experience at MIT proved to be invaluable to Professor Bertero's experimental research activities conducted later at the University of California at Berkeley on the behavior of structures subjected to seismic excitations.

By the time Professor Bertero came to Berkeley in 1958, the interest in blast effects on structures in the USA had greatly diminished, leading him to shift his main interest to seismic effects on structures—an interest he already had developed as a result of the 1944 San Juan, Argentina earthquake. Other researchers, including myself, made a similar change during the 1950s. Professor Bertero's continuing experimental research at the University of California at Berkeley focused primarily on understanding the inelastic hysteretic behavior of both steel and reinforced concrete structural elements and systems subjected to large-deformation cyclic inputs expected to occur during strong earthquakes. The understanding he developed led him to become a specialist in defining analytical models for use in conducting seismic performance evaluations. Having also developed a full understanding of the theory and application of structural dynamics, he could then treat the entire structural problem, namely design, modeling and analysis, and assessment of seismic performance.

Professor Bertero has very much enjoyed the broad role of being a teacher, i.e., transferring knowledge to students in the classroom, guiding them in their individual research, presenting papers at conferences, giving seminars and special lectures to the engineering profession, serving on technical committees, and consulting with practicing engineers. He has often stressed the importance of treating the complete structural system in developing a sound seismically resistant design, including considering related architectural, geotechnical, and construction issues. Consistent with this approach, while Professor Bertero supports specialization in civil engineering degree programs, he strongly encourages those students to gain a breadth of knowledge in related fields. At issue is the time required for civil engineering students to complete their degree programs. In the interest of gaining both specialization and breadth of knowledge, perhaps there is once again a need to consider increasing the bachelor degree program to five years and/or increasing the master's degree program to two years. Presently, we rely considerably upon the hope that students will gain the needed breadth of knowledge after entering the engineering profession.

In closing, I express my admiration for the accomplishments of my good friend Vic Bertero over his long and distinguished career, accomplishments for which he has received numerous prestigious honors and awards. He is now recognized worldwide among his peers as a legend in the broad field of earthquake engineering. I wish him continuing success and good health.

Joseph Penzien
Professor Emeritus of Structural Engineering
University of California, Berkeley
July 2008

CONNECTIONS

The EERI Oral History Series

Vitelmo V. Bertero

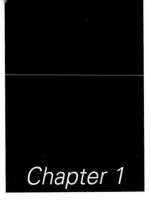

Growing Up in Argentina

When I was a little boy, I very much liked building little structures with my Meccano toy.

Bertero: I was born on a farm outside the small city of Esperanza, Argentina, in the Province of Santa Fe, on May 9, 1923.

My grandfather, Lorenzo Bertero, was born in Italy and had immigrated to Argentina. My father, Victorio Bertero, was a farmer like his father. My mother, whose maiden name was Lucía Gertrudis Risso, was the daughter of Juan Risso, who was born in Italy, in a village between Udino and Trieste, near the border with Yugoslavia, and of Teresa Dehrn, who was from Germany originally. Juan Risso immigrated to Argentina like my father's father. I had only one brother, Humberto Bertero. These relatives have all passed away now [2007].

When I was six years old, we moved into the city of Esperanza. In this region of Argentina, it was relatively easy to learn other languages besides Spanish, because its population was the result of the first agricultural colonization in Argentina in 1856 by a group of nearly

Figure 1. Schematic map of Argentina showing selected locations discussed in this oral history.

300 immigrants (called *colonos*) who came from Switzerland, Germany, France, Belgium, and Luxembourg. Ten years later, another group of colonos came from Italy, Poland, Syria, and Lebanon. The Santa Fe Province of Argentina had separate colonies, called Las Colonias. It was common that the people talked in German, Italian, and French, but not in English. Thus, although my education in Argentina was primarily in Spanish, in my eleven years of education in Esperanza I was also taught English, German, Italian, and French.

My school in Esperanza, *Colegio San José*, was run by a German Catholic order and so German was taught—and everything was taught very strictly!

Reitherman: Is Esperanza today still recognizably the same, still a city that is the center of the different European cultures that immigrated there years before?

Bertero: Yes. For example, in Esperanza still at present there exist two big buildings or facilities, one called *Sociedad de Canto* (Society of Singing) that used to be the German and Swiss Society. The other is called the *Sociedad Italiana* (Italian Society). These were large salons that were used (and still are used) by the descendants of the German and Swiss immigrants and Italian immigrants to celebrate their native festivals— that is, their days of religious and/or historical significance. During World War II, the different European colonies had different ties back to their home countries. Sometimes, the Germans would go to a festival and at the end when they tried to leave they found that the tires of their cars were cut.

Reitherman: What language was spoken at home?

Bertero: My father spoke Spanish and Piedmontese, an Italian dialect. Piedmont is in the northwest corner of Italy, where the Alps form Italy's border with France and Switzerland. The capital of the Piedmont region is Torino, or Turin in English. In Italy, there are several dialects—and there are certainly several types of Italian cooking! When I went to Italy for the first time as an adult, I tried to find in the restaurants a special Italian dish, *bagna cauda*, something like a fondue, which you can tell from its literal meaning in Italian: "warm bath." It is made with heavy cream, olive oil, garlic, and anchovies. You dip vegetables and various things in it and eat it steaming hot—and you need to drink wine along with that. My grandmother made it often in the wintertime. But it was not an Italian dish made all over Italy. It was only a recipe in the Piedmont region. Wherever I was in Italy—Roma, Venezia, and so on—I looked for this on the menu, but no one knew what I was talking about. I asked a professor friend in Italy, and he said, "Oh, you want to eat bagna cauda? You can have it at my house. I will call my wife's mother, who is from Turin, and she will know how to make it, because the only place they eat bagna cauda is in the Piedmont." And as an immigrant to South America, my Italian grandmother, the mother of my father, brought that Piedmont tradition with her. It is remarkable how people carry their traditions with them.

Going to School

Reitherman: What was your primary or elementary school like?

Bertero: I started elementary school, grades one through six, in 1929. At that time, when we went to elementary school, we wore a gray uniform with a bow tie. We had one basic teacher each year in a classroom with about twenty-five students. In high school, we had a separate teacher for chemistry, physics, mathematics, history, geography, and for each of the languages. Spanish was the basic language used for instructing the students.

In the elementary school, the subject that was so very difficult for me was penmanship, calligraphy. You had a special pen to dip into the ink, and you had to write using Gothic letters. The teachers, most of whom were priests and brothers of the Catholic order of the *Verbo Divino*, were very strict. And they taught us very much. I am very grateful for my education. I had very good teachers.

Reitherman: In these EERI oral histories, readers find it interesting to learn how a famous earthquake engineer first began to be interested in that subject. What class did you take in high school that was most related to your later earthquake engineering specialization?

Bertero: I do not think that I am famous, but I will answer your question. It has to be noted that my interest in earthquake engineering came later, during my university studies. However, my interest in engineering started even before my studies in high school, which I can tell you about. We had six years in the elementary school of the Colegio San José, followed by five years in the high school. In high school, the subjects, or courses, that were the most interesting to me were physics and mathematics. We had three years of physics in high school. I liked physics. I had good teachers in the courses of physics.

Reitherman: Not just one year, but three years of physics!

Bertero: However, there were other events that happened before then during my elementary education that began to give me an interest in what I would do later in life, that is, to become an engineer. The first occurred in 1928-1929, when I started primary school, with the construction of the house that was to be the home of my family in the city of Esperanza. The family house is now a school for teaching English. It was designed by a young brother of my mother who a few years before 1928 had gotten a degree of technical constructor that allowed him to design and construct buildings up to three stories in height. He supervised all the construction of the house, its separate garages, big birdcages, and the cisterns for storing rainwater to be used for the garden. I was attracted to the kind of work that he was doing and enjoyed hearing his conversations with the bricklayers.

Early in the 1930s, the construction of a new Catholic parish church started, which would replace the old, smaller one. Another uncle of mine was in charge of this new construction. The construction of this new church really attracted me, particularly the two high slender towers. The church remains a landmark building in Esperanza today—I just saw it again recently, in September 2006, when I visited Esperanza. It was a nice trip, when the city was celebrating the 150th anniversary of its founding. I met with the mayor and they declared me a Distinguished Citizen (*Ciudadano Ilustre*) and provided me a place of honor to view the parade.

Due to the above reasons, in my childhood I became attracted to the construction of real structures. And my parents at this time bought me a toy set of little steel elements that allowed me to construct models of structures. It was called Meccano. When I was a little boy, I very much liked to build little structures with my Meccano toy.

Reitherman: When I was a kid we had a construction toy called an Erector Set, which I think was an American product similar to the Meccano line made in England. Your Meccano toy sets came with all the little parts made of metal—struts, tiny clip angles and gusset plates, bolts, nuts, pulleys?

Bertero: Yes. The Meccano parts were all steel, no plastic. You could build anything with such a toy set. I liked to build bridges, and I liked to build towers. I started to learn how to construct models of different types of structures and enjoyed finding out the most efficient way of constructing each of those models. I was able to build a model of a railroad system that crossed over a bridge. Each year at Christmas my parents would give me a supplement to my Meccano set to build something new. That had a big influence on me. At that time in Argentina when I was growing up, it was not like it is at present in the U.S., when we have vocational counselors in schools letting the students know about the different professions or occupations. In most cases, the parents chose the student's future. For example, my mother wanted me to be a medical doctor.

Reitherman: You must have been in high school in 1939 when World War II began. Do you remember how you first learned that the war was starting? Newspaper headlines and articles? Listening to the radio?

Bertero: In a way, I learned about World War II before it happened. I had an uncle who was a technician who went to Germany to inspect ferryboats and other means of transportation one year before the war started. He came back to Argentina, and he told my family that Germany was preparing for war. He said that he would go to any industry, and he would see engineering work being done to be able to produce armaments and other products needed by the military, which he interpreted as preparation for war.

At that time, the Argentine military leadership had been trained in Germany, and most of them were sympathetic with Germany. This created some problems, as soon as World War II started.

In the Atlantic Ocean off Brazil, the German Navy sank a British ship.[1] This triggered a serious conflict between Brazil and Argentina.

Brazil was allied with the side of the war that the USA and Britain were on, and airfields were set up by the Americans to be able to fly the relatively short distance from there across the Atlantic to North Africa. Paraguay leaned toward the Axis powers. Argentina had its military and cultural ties to Italy and especially Germany, but it had the policy of maintaining its neutrality. The connections of European-colonized South American nations with either the Allied or Axis side caused regional tensions.

I always have thought that this political and governmental disagreement between Brazil and Argentina was the main reason for the interruption at the end of 1942 of my university studies, which I had started in 1940 after I completed my schooling at the *Colegio San José* in Esperanza.

Reitherman: To summarize, in 1939 you finished your first eleven years of school, which had included six years at your primary school, followed by five years at your high school. This brings the account of your early years up to the point where you became a college student.

---

1. On November 23, 1942, a German U-boat sank the cargo ship *Ben Lomond* in the south Atlantic. Four-fifths of the crew members died, and one, Poon Lim, survived 133 days alone in a life raft, still the record for survival adrift at sea, before floating to safety at the Brazilian coast.

Chapter 2

A Student at the University

When I started in civil engineering, there were about a hundred and twenty students registered. Only about twenty-five of these students graduated as civil engineers.

Bertero: I did my university studies toward the degree of civil engineering at the *Facultad de Ciencias, Matemáticas, Físico-Químicas y Naturales Aplicadas a la Industria* of the *Universidad Nacional del Litoral.* In other words, it was the faculty of sciences, mathematics, physics, and chemistry, along with the technical and engineering fields, as applied to industries. The university had the name *Litoral,* because its campuses were located along a river—the Paraná River, a major waterway in Argentina.

At that time in the 1940s there were six major national universities in Argentina: Buenos Aires, Córdoba, Cuyo, La Plata, Tucumán, and Litoral. Each of these universities had a faculty where you could study civil engineering, but I chose the faculty of the *Universidad Nacional del Litoral,* located in the city of Rosario, because it was the closest to my parents' home in Esperanza, and because I was already somewhat familiar with the civil engineering program at this university where a cousin was already in his fourth year of his studies for that degree. To get the degree of civil engineering normally required six years of study at the university. After three years you could get a degree as a surveyor, but six years were required to receive an undergraduate diploma in civil engineering.

Studying Civil Engineering

Reitherman: A civil engineering department in the USA today is almost always called a civil and environmental engineering department. There are a few department-wide courses expected of all the students with that major in their first year or two, but then there is significant concentration in one sub-discipline. Is it fair to say that today's civil engineering students tend to be directed into a specialized aspect of civil engineering—structures, geotechnical engineering, transportation, and so on—as compared to more of a generalist approach to civil engineering education when you went to college?

Bertero: Yes, that is true. In Argentina at that time, there were none of these divisions within the faculty of civil engineering. One problem in civil engineering education today, and the education of professionals in general, is its specialization in just one discipline.

Reitherman: Let me ask you about architecture and engineering, because you are known to be a strong advocate of considering all the various aspects of a construction project as a whole, not just considering structural engi-

neering apart from other disciplines. What was it like in the civil engineering school at the university at Rosario? Did the engineering students take some architecture classes? Did the architecture students take some engineering classes?

Bertero: Civil engineering students had to take three architecture courses. Architects had to take statics, strength of materials, and design, such as a design course in concrete and masonry, and another in the design of structures made of wood and metal (aluminum, but particularly steel).

Most of the professors teaching engineering courses in those days at my university devoted a majority of their time to their careers as practicing engineers rather than teaching. They were not doing any research, and that was a weakness. There was practically no communication outside the lecture hall between student and teacher.

When I started in civil engineering, there were about one hundred and twenty students registered. Only about twenty-five of these students graduated as civil engineers. We had to take six courses in the first year. About half the civil engineering class was gone after the first year. However, because of politics, students could stay in the university—even for fifteen years!—even though they were not progressing. Most of those long-term students devoted their time to student politics rather than study. But if you liked to study and work hard, I think in general that the program in civil engineering and the material presented by the professors, particularly those teaching mathematics and courses related to structural engineering, was very good.

Teaching Styles of the Professors

Bertero: Professor van Wyck, originally an engineer in Holland, taught me statics, strength of materials, and stability. He had a heavy accent.

Some of the women students, who were enrolled in architecture and were taking statics, would occasionally laugh at his accent, and he would stop and ask, "Can you tell me why you are laughing?" He was an excellent teacher and very serious about his teaching. The material that he presented in the above three courses was very useful to me.

Reitherman: What was the method of teaching? Was everything written on the blackboard?

Bertero: To answer this question I need first to clarify that most of the courses in civil engineering consisted of lectures presenting the theory that was involved, lectures given by the professor in charge of the course, and then laboratory sessions in which we carried out solutions of practical problems under the supervision of the adjunct or assistant professor of the course. The material that was covered in the theoretical lectures was written on the blackboard. One of the professors from Buenos Aires who taught electrical engineering would come to the class, take his watch off, and start teaching and writing on the blackboard until he noted that the time was over. When the lecture was done, you would not see the professor again until he lectured again. You never actually met with the professor. Because all of the teaching was with lectures, and because the professors did not do research, the material that they presented each year was the same.

Professor van Wyck was different. You could go to see him in his professional office for an hour or more. There were some of these professors, but they were the exception. In other words, the teaching system was completely different than that which I was to experience later on as a graduate student at MIT and also different than what we use at the University at California at Berkeley, particularly in teaching graduate courses.

Reitherman: Here, where we're sitting for this interview, in your office at the Earthquake Engineering Research Center at U.C. Berkeley's research facility in Richmond, California, you have a "blackboard," but it's actually a "greenboard." But when you say "blackboard," you mean the black-colored boards made of slate?

Bertero: Yes. And when the professor finished the lecture, in would come an employee whose job was to wipe off the blackboard. You could say that the basic instructional technology of the day was the blackboard and a piece of chalk—not photographic slides, not overhead projector, not computer projection.

Reitherman: Did you have one basic textbook for each course? Did you have one basic book for statics, for example?

Bertero: In general, no. There were some exceptions, like for physics, structural analysis, and design of reinforced concrete structures. Most of the professors would give a list of references during their first lecture. There was one reference that was given by several of the professors, and it was the German refer-

9

ence book, the *Hütte Manual of Engineering*.[2] In 1938 it was translated into Spanish. A large number of students bought this edition. I still have mine. This book, even today, is an excellent resource for the engineer.

As the same professor would teach the same course each year without introducing any important change in the material, the student center had compiled the notes taken by good students and reproduced and sold them to the future students. As a consequence of this system of teaching, not all the students attended all the lectures delivered by the professor. All the students had to take an examination in each course. The exams were given in December, March, and July. For most of the courses, the examination consisted of an oral exam. However, there were some courses that required written and oral exams. In the oral examination, usually the student had to answer a question on a subject that was named on a slip of paper

that the student would draw randomly from a container, a *bolillero*. The student had to go to the blackboard to solve the problems or answer the questions formulated by the professor that taught the course and by two other professors.

Reitherman: Having the student go to the blackboard to diagram or work an equation to get the answer to a question sounds as difficult and stressful as the typical graduate degree oral exam in the U.S. And you say this was required of the undergraduate students?

Bertero: Yes. And as I said, most of my initial classmates did not finish the program.

The grade that the student could receive based on his or her performance on these examinations ranged from high to low on this scale: *Sobresaliente* (outstanding), 10; *Distinguido* (distinguished), 7 1/2; *Bueno* (good), 6; *Aprobado* (passing grade), 4; and a failing grade of *Bochado* or *Reprobado*.

---

2. *Enciclopedia del Ingeniero y del Arquitecto Compilada por la Academia Hütte de Berlin.* The Hütte academic society of Berlin (Akademischer Verein Hütte) published the first edition of this encyclopedic engineering manual in 1857 and still maintains the reference work in updated editions. It includes sections on math, physics, and chemistry, as well as civil, mechanical, electrical, electronic, and other engineering and technology disciplines. The 32nd edition in German was published as *Hütte: Das Ingeniur, Wissen*, Horst Czichos et al., editors, Springer-Verlag, Berlin, 2004.

Chapter 3

In the Military During World War II

Later, in 1949, that girl was to be my wife—after the war and after I had finished my undergraduate education and had a job.

Bertero: At the end of 1941, partly because of the problems that developed between Brazil and Argentina that related to World War II, I was recruited into the military to serve as a soldier for three months at the artillery regiment located in Campo Mayo in the Province of Buenos Aires. In other words, I was drafted into the army. At that point, I had passed all the examinations required in the second year of my civil engineering studies.

The camp near Buenos Aires was hot, about 95 degrees Fahrenheit. They would have us run through a field filled with short trees, bushes with sharp stickers, and make us fall on them. The middle level officers in the military in charge of the new soldiers did not particularly like the university students. The students didn't know how to ride a horse, but they would just make them get on and ride.

I was born on a farm and knew how to tie the girth on the

horse, but the other ones did not, and as soon as they got on the saddle would slip, and the horse would start acting crazy. The sergeant would be mean with the young soldiers.

Reitherman: What did the military train you to do on horseback?

Bertero: At that time, the artillery pieces were pulled by the horses. You mounted the horses and directed them to transport and maneuver the artillery pieces.

In April of 1942, I returned to Rosario to continue with my university studies, but at the end of this year and before I finished with all the required examinations of the six courses that I took, I was drafted into the Argentine Army as a second lieutenant of artillery, and I was assigned to serve at a regimental base that was under construction near the city of Goya in the Province of Corrientes. This province is located in the north of Argentina between the Paraná River, on the west, and with Brazil on the east. As I have described, there was tension between the two countries because of World War II.

Meeting Nydia, Wife-to-Be

Bertero: After initial training in the city of Paraná as a second lieutenant of artillery, I arrived at the city of Goya in the Province of Corrientes, which is on the Paraná River. About twelve of us young officers didn't have any military quarters to stay in, and they sent us to an old hotel in the city. It was not very convenient. We went to the officer in charge of the regiment requesting to have our own clubhouse or *casino* as it is called in Spanish. He said that if we could find a house to rent,

the military would pay for it. And then he said to me: "Since you are studying engineering, you're the one who should find the house." So I was the one who went out looking for a house to rent. I looked for about two weeks and could not find anything. Then I was told that there was an elderly lady, who was the mother of thirteen children who had now grown up, who lived in a large house that was getting to be too much for her.

After making an appointment for inspecting the house, I went to see it and I met the elderly lady, Doña Felipa Vilas, along with one of the lady's grownup daughters, Ana Vilas. There was also visiting there that day one young granddaughter of about fifteen years. Her name was Nydia Ana Barceló Vilas, and her parents were Adolfo Barceló and María Vilas. Later, in 1949, that girl was to be my wife—after the war and after I had finished my undergraduate education and had a job. We had a nice fiftieth anniversary celebration in 1999, and now, in 2006 [when this particular interview session occurred], we have been married fifty-seven years.

So, we young officers rented the house. The grandmother and her daughter moved to a smaller house beside the big house, and the bachelor officers moved into the big house. It was a big, old, typical Argentine house, with the rooms built around a large courtyard.

Near the end of 1943 I filled out an application requesting to be released from my duties in the Army so I could continue my studies in civil engineering at the university. The Chief Officer, who was in charge of the Artillery Group, after reviewing my application considered that my duties with the fatherland ("*la Patria*") were more important than my studies, and therefore he threw away my application.

Reitherman: You mean he literally threw it away?

Bertero: He crumpled it up and threw it in his wastebasket.

Recovering From Injury

Bertero: Later in 1943 I was kicked by a horse in the leg and was injured. To ride from our rented house to the city, we rode on horses. One of the assistants who brought the horses did not handle the horses properly, and one horse kicked me. I was injured and in pain, and my leg quickly swelled so that my boot would not come off and had to be removed by cutting it. The x-ray machine in Goya was not working well enough to detect if my leg was fractured. They told me to take the boat down the Paraná River to the military hospital in the city of Paraná, but that would take two days. I preferred to go to see my doctor in Rosario by crossing the river on a ferry boat, then taking the bus to Rosario, which would only take about one day, so that is what I did. When I arrived in Rosario my doctor took me to the hospital to do surgery on my leg.

After I recuperated from my injury, sometime later the military let me go back to the university.

Chapter 4

Back in College, Becoming a Civil Engineer

It was the 1944 San Juan earthquake that recruited me into the earthquake engineering field.

Reitherman: It is now 1944. You are back in the university finishing your undergraduate civil engineering degree. But I don't think you have described anything yet in your life that indicates why you entered the earthquake engineering field. How did that happen, Professor Bertero?

"Earthquake-Resistant Construction" and "Earthquake Engineering"

Bertero: First of all, in 1944 the term "earthquake engineering" was not yet used officially. There was some literature and engineering practice regarding the problems created by earthquakes, but it was usually called "earthquake-resistant construction" and it was really more about construction than engineering. The ground motion and structural response aspects were usually called engineering seismology. The Earthquake Engineering Research Institute did not exist yet, being only organized in 1948 and beginning to hold meetings and function in 1949. In fact, the familiar EERI almost ended up as SERI, Seismic Engineering Research Institute. Three of the founders, John Blume, George Housner, and R. R. Martel, thought it should have "earthquake" in its name and so it is the Earthquake Engineering Research Institute and "earthquake engineering" has become a common term in significant part because of EERI. EERI was an outgrowth of the Advisory Committee on Engineering Seismology at the Coast and Geodetic Survey. The application of more advanced engineering analysis methods for seismic design was just beginning.

In 1952 when EERI held one of its first big functions, a conference held at UCLA in Los Angeles,[3] it decided to include both the Cold War subject of blast resistance and defense against nuclear attack along with the earthquake subject. The organizers weren't sure they would get a big enough turnout of engineers and seismologists if the conference was just on earthquakes. The first of the World Conferences on Earthquake Engineering, with that now-familiar name "earthquake engineering" in its title, didn't occur until 1956. So, as of the mid 1940s, there really wasn't a well-defined earthquake engineering field a student could enter, the way there is today.

Reitherman: I note from the front matter of those 1952 proceedings that to get a copy

one would write to Karl Steinbrugge at his San Francisco office address, the same way a few years later one ordered the 1956 World Conference proceedings, with the order filled from boxes of the volumes Karl had stacked under a table in his office.

Bertero: Yes, it was different back then. You can see that back in 1944, I had not yet any opportunity to study engineering as applied to the earthquake problem. We can also discuss how there were even some additional problems in Argentina in the universities in obtaining information from other countries where there were more earthquake studies underway. I have told you how earlier events in my life—my uncle's building the church in Esperanza, my Meccano toy sets, my interest in physics, majoring in civil engineering—helped to prepare me for a career in earthquake engineering, even though I did not think at an early age that this was to be my career. But one event stands out. It was the 1944 San Juan earthquake that recruited me into the earthquake engineering field.

San Juan Earthquake, 1944

Bertero: When I was back from my military service to finish my civil engineering education, the January 15, 1944 San Juan earthquake occurred in western Argentina. It is still the largest natural disaster in the history of Argentina. It caused about 10,000 fatalities. It was a magnitude 7 earthquake, and unlike some of the earthquakes that occur at great depth in Argentina, well inland from the Pacific Ocean subduction zone, the 1944 earthquake was shallow and especially damaging.

---

3. C. Martin Duke and Morris Feigen, editors, *Symposium on Earthquake and Blast Effects on Structures*. Earthquake Engineering Research Institute, Oakland, California, 1952.

Two professors at the *Universidad Nacional del Litoral*, Dr. Alfredo Castellanos and Dr. Pierina Pasotti, both earth scientists, studied the disaster in San Juan. Each of them in less than two and a half months published a report on the studies that they conducted. They were my professors in 1942 in a course called *Fisiografía, Mineralogía y Petrografía*—basically a geology course. There were no courses in civil engineering on earthquakes in Argentina at this time, and I don't know of any that were offered in other countries yet either. Castellanos offered to start a course on what he wanted to call *Sismología Pura y Sismología Edilicia*, or pure seismology and seismology applied to buildings. It was intended for the civil engineers and architects, but there was not enough support. I met with him individually and read his report on the earthquake.

The tremendous damage the earthquake caused and the reports of Professors Castellanos and Pasotti brought me into the field that would later become earthquake engineering. Castellanos basically wrote what we would call today the reconnaissance report on the earthquake.[4] Professor Castellanos, by the way, started out to be a doctor of medicine. You recall that I described the traditional system where the parents determined the children's occupations. He graduated from medical school because his parents wanted him to. Then, he said, "I'm not

going to be a doctor, I want to be a geologist," and he went on to do that. His dedication to research and to teaching had a very big effect on me. Unfortunately, despite the fact that Professors Castellanos and Pasotti emphasized in their reports the urgent need for education of the students in civil engineering and in architecture, there was no support to do that. Furthermore, at that time, during World War II and right after, it was practically impossible to acquire foreign publications and new books, so it was not easy to learn about *sismología edilicia*.

San Juan is near several faults, and the historical pattern was that they moved the town after an earthquake, but they just moved it farther from one fault and closer to another. One national government building in San Juan performed well, and its construction was studied to see why. What was learned from this study helped to develop prescriptive requirements regarding materials to be used, workmanship, and particularly requirements on proper detailing, which improved earthquake-resistant construction. These requirements initially helped the builders improve the earthquake resistance of the city. Later on, real earthquake engineering techniques that the engineers could use were developed and implemented by simple code provisions.

Reitherman: When was the first time you were in San Juan and saw the destruction first hand?

Bertero: It was in 1945, although I was already somewhat familiar with the destruction from the report of Dr. Castellanos and discussion with him at the university. The reconstruction was just starting. A general, named Plácido Vilas, an uncle of the girl who was to be

---

4.   Castellanos, A. *Anotaciones Preliminares con Motivo de una Visita a la Ciudad de San Juán a Propósito del Terremoto del 15 de Enero de 1944. Publicación del Instituto de Facultad de Ciencias Matemáticas, Físico-Químicas, y Naturales Aplicadas a la Industria* of the *Universidad Nacional del Litoral*, Santa Fé, Argentina, 1944.

my wife, had been appointed as provisional or temporary governor of the San Juan Province.

They really did a tremendous job in the reconstruction, applying temporary seismic code provisions. It was not a very sophisticated code, but it contained very specific and precise details for the construction. This code was imposed under rigorous control. Everyone had to comply with the regulations. No more adobe or unreinforced concrete. This was very fortunate because when the 1977 Caucete earthquake happened, which had a magnitude of 7.4, no one was killed in the reconstructed city of San Juan.

I consider that the San Juan earthquake in 1944 really was the initiation of earthquake engineering in Argentina. The Chilean building code seismic provisions came earlier and the Chileans were more advanced. They had not only suffered more earthquakes, their earthquakes were more destructive. The Nazca subduction plate releases shallower earthquakes under Chile as it dips and moves eastward under the South American continent. By the time the Nazca plate is under Argentina, it usually releases earthquakes at a greater depth, and so the earthquakes are farther from the construction at the surface and the ground motions are weaker.

Earthquake Engineering Developments In Other Countries

Reitherman: In August, 1906 Chile had a larger and equally devastating earthquake near Valparaíso as compared to the one that happened in northern California in April of that

year.[5] That earthquake near Valparaíso seems to have initiated the interest in modern earthquake engineering and seismology in Chile. The Comte de Montessus de Ballore, one of Europe's leading seismologists, left France to set up Chile's national government seismographic program and also started university classes on earthquakes. So that gave Chile an early start in the field. But it wasn't until their 1939 Concepción earthquake that Chile actually adopted a seismic code.

What do you think of the historical pattern in some countries where one or more big earthquake disasters set in motion the development of that country's earthquake engineering? That pattern seems to apply to the 1944 San Juan earthquake you describe, the 1906 earthquakes in Chile and California, the 1931 Hawke's Bay earthquake in New Zealand, and the 1931 and 1935 earthquakes in Baluchistan that launched the seismic code of India.

Bertero: Your observations are correct, however, one country with a large earthquake problem that did not progress till very late was China. China historically had many, many earthquakes and recognized the problem. But its interest in modern earthquake engineering was initiated relatively recently. That is a strange situation. The same thing has happened in some other countries, of course, where they have had many earthquakes, but have only recently seriously begun their inter-

---

5.   The August 17, 1906 earthquake in Chile had a moment magnitude of 8.2. Just as in the San Francisco, California earthquake of April 18 of that same year, Valparaíso suffered as much or more from fire as from the initial earthquake damage.

est in research and practice in earthquake engineering programs, or maybe have not even started that.

Reitherman: The Chinese seem to recognize Dr. Liu Huixian as their original earthquake engineer. He began the seismic code development and research program at Harbin Institute of Technology in 1954. That did not coincide with any particular earthquake disas-ter in China. Instead, the government recognized the need for a seismic construction code, and Dr. Liu was put in charge.

Bertero: Different countries took different paths. We can talk more about how all the aspects of society should be involved in solving earthquake problems. The engineers cannot be expected to do it alone.

Chapter 5

Graduate Student at MIT

Beside me was seated a lady from Chile. It was also her first time on an airplane. As the airplane was shaking in the storm, this woman was praying on her knee with a rosary.

Reitherman: When you finished your undergraduate civil engineering degree in 1947, did you decide to go to graduate school right away?

Bertero: No, not right away. I was interested in conducting research and doing some consulting professional work on the design of civil engineering structures considering their nonlinear inelastic (plastic) behavior, particularly when they could be subjected to the effects of severe earthquake ground motions. It is for this reason that early in 1946, before receiving my civil engineering degree, I accepted a position of research assistant at the *Instituto de Estabilidad*, an institute of structural stability, of the Faculty in Rosario. I started to carry out experimental and analytical research on the mechanical behavior of structural elements in their linear and nonlinear elastic and particularly their inelastic (or plastic) regions.

Today, nonlinear elastic analysis is important if you want to use a technique like prestressing to allow concrete members to rotate at joints and be re-centered so that there is no residual deflection after the earthquake. As you allow rocking, the resistance changes as the geometry changes—it is nonlinear—even if none of the material behaves inelastically.

At that time, it was very difficult to get information such as technical reports and journals from countries that were doing research on these topics. The reports written by my former professors, Doctor Castellanos and Doctor Passoti, and the personal discussions that I had with them, helped me to understand the importance of engineering seismology and engineering geology in dealing with the complex engineering problems that can be created by significant earthquake ground motions. The director of the *Instituto de Estabilidad* was Professor Roberto Weder, who had received his degree of civil engineering from the Faculty in Rosario in the 1930s, and then in 1946 received his master's degree in civil engineering from the Massachusetts Institute of Technology. He helped me learn about the nonlinear inelastic behavior of materials. He was born in a German colony around the town of Humboldt in Argentina, which was close to Esperanza. Although he was encouraged by the MIT professors to continue his doctoral studies there, he decided to come back to Rosario.

As in 1947 my salary as a research assistant in the Faculty of Rosario was not enough to support myself, in early 1948 I started to work as a Technical Director of the *Talleres Metalúrgicos Angeloni*, or Angeloni's Metallurgical Factories. I was in charge of the design and then the direction of the fabrication and erection of civil engineering steel structures.

Marriage to Nydia Barceló Vilas

Bertero: In 1949 I married Nydia Ana Barceló Vilas. You will recall that I described meeting her six years before, the young granddaughter of the landlady of the house that I inspected for the military and that was then rented for the young military officers. Later in 1949, I was appointed Adjunct Professor (*Jefe de Trabajos Prácticos*) of Stability, teaching a course for civil engineering students at the *Instituto de Estabilidad*, and also teaching a course for architecture students on the design of wood and steel structures. In 1950, I joined Professor Roberto Weder in a structural consulting firm named Weder-Bertero. Between 1950 and 1953 we designed several reinforced concrete and steel building structures. As a consequence of this professional work, it became clear to both of us that there was a need to improve the current practice: first conducting research, and then incorporating the results into the teaching of the students at the universities.

Lack of Graduate Engineering Education in Argentina

Bertero: At that time in Argentina, there were no graduate schools in engineering. Only recently, about the beginning of the twenty-first century, some of the engineering faculties have offered graduate programs for the doctoral degree in engineering in Argentina.

At the University of Buenos Aires, the first civil engineer who received such a PhD degree was *ingeniero*, or engineer, Raúl Bertero, early in 2006. Raúl earned his master's degree at the University of California at Berkeley in 1992 and started working on his doctoral thesis early in 1997 under the guidance of the engineer Alberto Puppo, a professor of the University of Buenos Aires, and myself.

Reitherman:　I have come across Raúl Bertero's name before, and I always assumed he was a brother of yours, but you said earlier you had only one brother, Humberto.

Bertero:　Raúl has the same last name, and there is some distant relationship in the family tree going back to Italian ancestors, but we are not close relatives. However, the father of Raúl, Domingo Bertero, was like a brother to me and he also got his degree from the same Engineering Faculty, one year after I did.

In Argentina, back around the 1950s, laboratory facilities were not adequate or appropriate, and we didn't even have funds to obtain technical publications on what was being done in other countries. Beginning in 1942, the normal activities of the Argentine universities started to be disrupted by political and economic problems. The students started to organize protests that then were followed by strikes. The military government intervened in the running of the universities and some of the university authorities were removed. There were big budget cuts for the universities.

Reitherman:　This would have been during the time when Juan Perón was the president and a strong centralized force running Argentina. In researching the 1944 San Juan earth-

quake, this interesting seismic fact turned up: Juan Perón, then a colonel in the military, met Eva there when they were doing a fundraiser for earthquake victims. It also seems this disaster had a lot to do with strengthening Perónist political fortunes, because they were sympathetic to the plight of the poorer people in the region after the earthquake.

Bertero:　Soon after that, Juan and Eva Perón began their decade-long tenure as populist leaders. It was not a time when the universities were favored.

Near the end of 1952, I decided that to improve my knowledge by conducting research it would be better for me to go to some foreign universities or institutions that at that time were conducting research in the areas that interested me, namely these challenging problems about inelastic behavior and seismic design. From the personal discussions that I had with some engineers who had been involved with research in Europe and the USA, and in talking with Professor Weder, it became clear to me I had two choices, MIT or ETH.[6] I decided to go to MIT.

Reitherman:　What made up your mind?

Bertero:　I had an economic problem, because I was by that time married and had two children, a daughter, María Teresa, then my son, Eduardo Telmo. Thanks to correspondence between Roberto Weder and MIT professors, I was able to get a research assistant position at MIT to provide some income.

6.　*Eidgenössische Technische Hochschule* (Swiss Federal Institute of Technology), Zurich, Switzerland.

Leaving Argentina for MIT

Bertero: In July, 1953 I left Argentina for the Massachusetts Institute of Technology. I could not take my wife and children right away, because the U.S. authorities said that I had only a one-year research assistant position, and I needed to demonstrate that this position would last more than one year. I could not get a certificate from MIT in advance, but fortunately, as soon as I arrived at MIT the people there were very nice and prepared a certificate stating that I would work as a research assistant during nine months and then as a research engineer for the other three months for the time that I needed to finish my studies. In about two months, my wife was able to travel to Massachusetts with my two small children.

Reitherman: How did you travel from Argentina to Cambridge, Massachusetts? By ship or by airplane?

Bertero: By airplane. Oh, what a trip! It took two days. First we flew from Buenos Aires to Sao Paolo in Brazil. Then another stop at another city in Brazil. Then in Havana, where we spent half a day while they were checking the airplane for some problem. Then to Boston. It was the first time I had been in an airplane.

This was before jet airliners, so it was a propeller airplane. The weather over Brazil was very, very rough. Beside me was seated a lady from Chile. It was also her first time on an airplane. As the airplane was shaking in the storm, this woman was praying on her knee with a rosary.

Reitherman: You mean she literally got out of her seat during the turbulence and knelt down to pray?

Bertero: Oh yes, she was on the knee! I will never forget my first airplane flight. It was an adventure, in other words.

When I arrived at MIT, I started to work on my master's degree and do my research assistant job. I received my MS in civil engineering in 1955, and then continued for my ScD in 1957.

Reitherman: Let's note for the benefit of readers that at that time MIT granted a ScD, or doctor of science degree, in civil engineering, as compared to the PhD. The ScD and PhD are equivalent doctoral degrees. Today, there are only a few U.S. research universities that still grant both degrees, rather than just the PhD.

Did you begin to study earthquake engineering as a master's student?

Blast Engineering—Not Yet Earthquake Engineering

Bertero: No, I was not taught earthquake engineering—it was not taught as a separate subject then. But I was taught different aspects of engineering that prepared me for earthquake engineering. At that time there were no funds for earthquake engineering research. All the research efforts were devoted to the effects of atomic (fission) bombs, or "A-bombs," and later the hydrogen (fusion) bombs, or "H-bombs." The goal was to develop reliable design and construction methods for resisting these effects on structures such as bomb shelters. This research was supported by the U.S. Army,

U.S. Navy, and the Federal Civil Defense Administration. At MIT, professors Bob Hansen, Myle Holley, Jr., and John Biggs[7] were doing research in cooperation with a group at the University of Illinois headed by Professors Nathan (Nate) Newmark and William (Bill) Hall.

The results of this research on structural dynamics were summarized in a two-week short course that was offered during the summer session of MIT in 1956. At this time, there was money from the U.S. civil defense agency to teach engineers how to design nuclear war shelters, underground shelters that would be protected from radiation and dynamic pressures in the atmosphere that an aboveground building would receive. But underground structures could still receive large blast loadings propagated through the ground.

During this course, a professor from Japan accomplished a very good English language summary about earthquakes and particularly about their dynamic effects on structures, and then he compared the state of the practice in the U.S. and Japan considering their seismic design provisions. This Japanese expert was the father of a young, bright student who later in the 1960s did graduate study at Berkeley. The father I am speaking about was later appointed the Secretary of the International Association of Earthquake Engineering.

Reitherman: That must have been John Minami.

Bertero: Yes, that was his name. He was a professor at Waseda University who came over in 1956. The lecture notes for the course were later the basis for what was possibly the first book that I know of written for the structural engineer designing civil engineering structures.[8] Norris, Wilbur, Holley, and Biggs would mention earthquakes from time to time, but at that time the Japanese were more advanced in earthquake engineering. Minami was an excellent teacher.

Professors at MIT

Reitherman: Tell me about some of the professors you knew at MIT.

Bertero: At MIT, I had very good professors, excellent professors. In the classroom, people like Charles Norris and Jacob P. den Hartog, for example, were tremendous. Other good professors were John Wilbur, Walter

---

7.   These three MIT professors later formed the consulting engineering firm Hansen, Holley, and Biggs. J. M. Biggs authored *An Introduction to Structural Dynamics*, McGraw Hill, New York, 1964, an early textbook in this subject area. Hansen's first research assistant was Joseph Penzien, who started in January 1948 in his graduate work to do research on blast effects on reinforced concrete beams. (*Connections: The EERI Oral History Series—Joseph Penzien*, Stanley Scott and Robert Reitherman interviewers. EERI, Oakland, California, 2004, p. 15). M. J. Holley, Jr. was later to head the Structures Division of the Civil Engineering Department at MIT.

8.   C. H. Norris, R. J. Hansen, M. J. Holley, J. M. Biggs, S. Namyet, and John V. Minami, *Structural Design for Dynamic Loads*. McGraw Hill, New York, 1959. This was one of the first, possibly the first, text published to survey the subject of structural design for dynamic loads, written for the civil or structural engineer, as distinct from works treating the mechanical or aeronautical engineering aspects of dynamics.

Maxwell Fife, Myle Holley, John Biggs. Wilbur and Fife wrote a structural engineering textbook together.[9] Then, some of them were perhaps not so outstanding in the classroom, but when you visited them in their offices, and they were always willing to meet with you, you learned more there. You remember I said that in Argentina some of the professors were not available outside the classroom. It was different at MIT, and I am indebted to my MIT professors.

I worked hard, and there was some economic pressure on my family. When I started my doctoral studies, my wife and I had two children, María Teresa and Edward. Two more were born while we were at MIT: Robert and Mary Rita. The five years I spent at MIT, getting my master's and doctorate and working another year for them, was an economic sacrifice for my family. It was also the time when I really prepared myself to become an earthquake engineer.

Robert Hansen

Reitherman: What about your doctoral advisor, Robert Hansen? Because so many people in the earthquake engineering field are familiar with the similarly named person of a later generation with the "o" in his last name, Robert or Bob Hanson, let's make that distinction clear here.

Bertero: Professor Hansen was perhaps the youngest of the engineering faculty that taught me at MIT. He was very dynamic and very interested in blast-resistant design, and particularly in the development and use of specially

devised loading machines for experimental work. He was a specialist in blast engineering. As I was interested in experimental work, they assigned me to him.

Reitherman: What was Hansen like?

Bertero: Hansen was an excellent engineer. He excelled as a specialist in blast-resistant design as well as getting funds for conducting the research needed in this area.

Jack Benjamin

Bertero: Jack Benjamin[10] worked with Robert Hansen at MIT before I arrived there, working on blast loads on reinforced concrete shear walls. He did very important research on that topic. Benjamin later went to Stanford, where he kept doing research on concrete and masonry shear walls. Another professor at Stanford, Harry Williams, worked on that research, funded by the U.S. military in the 1950s.

When I became a member of the faculty at U.C. Berkeley, I paid attention to Professor Benja-

---

9. Walter Maxwell Fife and John Benson Wilbur, *Theory of Statically Indeterminate Structures*. McGraw-Hill, New York, 1937.

10. Jack R. Benjamin (1917-1998) started in structural engineering as an undergraduate and master's student at the University of Washington, working for a time under Professor F. B. Farquharson on the investigation of the 1940 collapse of the Tacoma Narrows Bridge. He received his Doctor of Science degree from MIT in 1942, was in the U.S. military in World War II, taught briefly at Rensselaer Polytechnic Institute afterward, and then, from 1948 to 1973, was on the faculty of Stanford University. He was known for his structural design instruction, using analysis concepts such as visualization of deflected shapes as aids to design, and authored *Statically Indeterminate Structures*, McGraw-Hill, New York, 1949.

min's work. I attended one summer short course that he and Dr. Allin Cornell offered on Probability for Civil Engineers. We were friends. I went to Stanford once in a while to talk with him. Professor Benjamin retired from Stanford in 1973 and formed Jack Benjamin Associates.

Reitherman: It's interesting that Benjamin is noted for, among other things, his early research and advocacy of the use of probability in civil engineering. Another Stanford figure prominent in that subject area whom you mention, C. Allin Cornell, taught at MIT for almost twenty years beginning in the 1960s, after you were there, and then went on to become a professor at Stanford. Cornell and Benjamin authored a still-authoritative book on probability for engineers.[11]

Bertero: Yes. Allin Cornell has been a leader in the application of probability in the field of earthquake engineering.

Howard Simpson, Werner Gumpertz, and Frank Heger

Bertero: Then at MIT I had two other young professors who advised and helped me in my research. During the first two years, it was Professor William J. LeMessurier, the next three years it was Professor Howard Simpson. Professor Simpson later started the engineering firm with two partners who were also assistant professors at MIT then, the firm that is still called Simpson, Gumpertz, and Heger. They set that firm up in the Boston area in

1956. I also knew Werner Gumpertz, and I was a good friend of Frank Heger.

Simpson's strong capabilities were more in doing research and in the consulting field, which he went into, rather than being a professor. Gumpertz was the administrator and manager of the three. Heger was quieter. He made a trip to San Francisco in the late 1980s to talk with me when that firm was going to set up an office in San Francisco. I was then a professor at Berkeley and he asked for some advice on whom to hire.

Hansen asked me to help Simpson on the design and construction of a machine with a capacity of 300 kips to do blast simulation testing. It had to be able to exert its force in one millisecond. In one one-thousandth of a second, the force would suddenly increase to a very high level. It would then decline suddenly also, and there was a small negative phase when the pressure was less than the original static pressure. There was no such device at that time, so Professor Simpson started from scratch. Another lab at MIT developed a servo-valve that was quite advanced for that time. The machine was so sensitive, any small thing could make it malfunction. The loading was like a single earthquake pulse, except that the graph of it would show it going from zero to a maximum value of 300 kips in 0.001 seconds, and then back to zero in a few milliseconds and to a slight amplitude the other way, and then to zero after some milliseconds.

We had to use a special silicone hydraulic oil, because the heat generated was so great that ordinary fluid would catch fire.

The machine was completed, so we tried it out. The laboratory was in the basement.

---

11.   Jack R. Benjamin and C. Allin Cornell, *Probability, Statistics, and Decision for Civil Engineers*. McGraw-Hill, New York, 1970.

When we ran the test, it vibrated, and soon people from the third story came running in. Professor Norris was alarmed and said, "What is going on!" What was going on was a problem with that big machine, and we had to learn how to keep it from destroying itself and shaking the building.

Reitherman: It seems like your first earthquake engineering experiment, in effect, was to shake a building at MIT, by accident, using a blast load simulator, alarming the chair of the Civil Engineering Department in the process.

Bertero: It took a year to build the machine and get it functioning properly to test shear walls, and then I spent one year after my PhD to help to continue using the machine to conduct research. I was asked to assist the doctoral students who were then doing research with the testing machine.

Reitherman: How were the reinforced concrete shear walls loaded?

Bertero: The piston with a head on it to spread out its force pushed the top of a shear wall that was attached to the foundation of the building. There were many accumulators for the hydraulic power. The servo valve was the key to providing the load, which had to be applied in a thousandth of a second and then go back to zero to simulate the blast wave propagating through the structure.

The testing was needed to design underground nuclear shelters. The effects of a nuclear bomb at the surface, the blast and radiation, were so intense that the idea was to build underground shelters. Then the way the bomb affects the structure is to cause a blast wave to go through the soil and reach the structure. I was not very attracted to this whole subject. They were doing studies on the behavior of people that had to spend one or two weeks underground in these shelters. It is one thing to worry about natural disasters, such as an earthquake, and another thing to worry about hazards like nuclear attack that people are creating.

John Wilbur

Reitherman: You mentioned the Fife and Wilbur book, which was on indeterminate structures. What about the book Wilbur wrote with Charles Norris,[12] a structural analysis textbook that has stayed in print a long time?

Bertero: In my first semester at MIT, I took a course in structural design that Professor Wilbur taught. He was very good, very practically oriented. Professor Wilbur was very good at design.

Charles Norris

Bertero: In my second semester at MIT, I took a course on mechanics and dynamics that Professor Norris offered. He was so clear, in and out of the classroom, that I consider him as the best teacher that I had.

During the third year at MIT, as a candidate for the ScD degree, each week I had to participate in a meeting with Professor Norris where one of the students had to do a research presentation, covering progress and problems. All the

---

12. John Benson Wilbur and Charles Head Norris, *Elementary Structural Analysis*. McGraw-Hill, New York, 1948, first edition; fourth edition 1991.

doctoral students had to do that. This was a good experience for me. Thus, when I arrived at U.C. Berkeley, I required this for all graduate students doing research under my supervision.

I believe that Professor Norris left MIT because he did not believe in the emphasis on computers that was becoming prominent there. He went to the University of Washington, where he had been a student, and he taught there for many years, and was the chair of the Civil Engineering Department and later Dean of the School of Engineering. That emphasis in the mid-1950s on computer science at MIT was also one of the things that led me to seek another university when I finished my doctorate.

Myle Holley, Jr.

Bertero: When I started doing my research, I worked closely with other professors also. First came Myle Holley, Jr., because I was doing work with reinforced concrete.

Professor Holley didn't have the charisma of Professor Norris as a teacher, maybe, but he knew his subject. When Hansen, Biggs, and Holley started their consulting firm of that name, they also teamed up with Professor Nathan Newmark at Illinois. Those were the two universities with the most talent in dynamics, plasticity, probability—subjects needed to do blast engineering in the Cold War and also to work on nuclear power plant issues like earthquakes.

I never will forget what Professor Holley said in his first lecture in his reinforced concrete class. He said, "Ductility is a blessing given to the structural designer." He was right. Why? Because the ductility of your structure can

overcome the great uncertainties that are involved in the modeling and analysis of the real performance of the designed and constructed structure. I have repeated this statement in my teaching, and I add the caution that you should not abuse the use of higher ductility to reduce the needed strength of the structure, because ductile behavior is associated with damage. Furthermore, it is necessary not to confuse physical ductility with the concept of a ductility ratio. I can talk more about this point later.

Other Professors at MIT

Bertero: Although I did not take the course that Professor Biggs offered, I had several discussions with him. Furthermore, I did some studies with him. The main one was not related to earthquake shaking, but was an experimental study on the vibration of a bridge from the traffic on it. I had to apply my learning about dynamics to that example of a bridge vibrating under traffic load.

Reitherman: As of the late 1950s, the term geotechnical engineering was not yet used?

Bertero: No, it was called soil mechanics. I took the course from Professor Don Taylor. After taking his course, I listened to the lectures, or audited the course, of Professors Karl Terzhagi and Arthur Casagrande, going to the other end of Cambridge, from MIT to Harvard.

My knowledge of soil mechanics at that time was practically zero. Professor Taylor was an excellent instructor and he had written a good textbook.[13]

---

13. Donald W. Taylor, *Fundamentals of Soil Mechanics.* John Wiley and Sons, New York, 1948.

As I worked on my doctoral degree in 1955, Professor Don Taylor passed away. Bob Whitman had left MIT for a couple of years for military service, and when he returned, he took over for Taylor. But Whitman was a structural engineer. He was a structural engineering student of Norris, but ended up working in the soil mechanics program at MIT. Whitman knew what the structural engineer had to know about the soil to design a structure, and that was a tremendous advantage. He has been very successful as a professor and researcher. He has contributed significantly to the advancement of earthquake engineering.

Reitherman: You have mentioned to me a couple of times that Jacob (Jappie) den Hartog was one of your best instructors at MIT. Tell me about him.

Bertero: He was a professor in the Department of Mechanical Engineering. He was from the Netherlands. I learned a lot from people in that department, also from the Division of Materials of the Civil Engineering Department. Den Hartog had examples that were so clear that even if you didn't understand the concepts when he started a lecture, you would really understand them when he had finished. He was a great teacher. He had written excellent textbooks on different engineering subjects.[14] If I learned about vibration well, it was because of his course. I took first the course of Charles Norris on mechanics of solids and dynamics.

Then, after the course with Professor den Hartog, I learned from Professor Stephen Crandall, who taught random vibrations and who had studied with den Hartog. It was dynamics combined with probability.

Joe Penzien, also from MIT, was on the University of California faculty before I was. When Professor Penzien had his first sabbatical, he spent it back at MIT, and he took the random vibration course from Stephen Crandall. Crandall was a very good teacher. Very theoretical, very fundamental. It was a mechanical engineering course, not a structural engineering course, but you learned the theory.

Reitherman: Looking up the biographical summary Stephen Crandall wrote about den Hartog,[15] we find some interesting facts. For example, Hartog got his start in dynamics and mechanical engineering with none other than Stephen Timoshenko,[16] the Ukrainian who left the Soviet Union after the Bolsheviks took over. Timoshenko ended up in the U.S. working for a while with Westinghouse, and den Hartog got his start there under him. Crandall says that Timoshenko turned the young den Hartog, an electrical engineer, into a mechanical engineer. Crandall tells the story of how Timoshenko referred a problem of machine

---

14. Jacob P. den Hartog, *Mechanics*, 1952, *Strength of Materials*, 1949; *Mechanical Vibrations*, 1947. McGraw-Hill, New York. All have been re-published in updated or re-printed editions.

15. Stephen Crandall, "Jacob Pieter den Hartog, Biographical Memoir," National Academy of Sciences, Washington, DC, 1989.

16. Stephen P. Timoshenko (1878-1972) wrote some of the classic works in engineering mechanics, elasticity, and strength of materials. He worked for Westinghouse Electric Corporation from 1923 to 1927, then joined the faculty of the University of Michigan. In 1936, he became a professor at Stanford University.

vibration to den Hartog. Resonant response of the shaft of a turbine was leading to fatigue failure. Den Hartog calculated that the necessary de-tuning could be accomplished by only a small variation, and so he boldly proposed using a shaft only 1/16 inch (1.5 millimeters) smaller in diameter, which solved the problem.

Another interesting sidelight is that Lydik Jacobsen, later to head up the vibration laboratory at Stanford and be a mentor to John Blume, was working there in the Westinghouse laboratory with Timoshenko in the early 1920s in Pennsylvania.

Bertero: I also took a class from the engineer who worked on the plastic House of the Future in Tomorrowland at Disneyland, which involved the Monsanto company. Albert Dietz was a materials expert, and he was interested in the new polymer materials. He taught a course on building construction and materials, but he was looking for new things. We are starting to use some synthetic materials today in structures, but it has taken a long time.

1956 World Conference on Earthquake Engineering

Reitherman: You were at MIT when the World Conference on Earthquake Engineering, the first one, was held at Berkeley in 1956.

Bertero: I couldn't go to it. It was not the official "First" World Conference. Professor Kiyoshi Muto came to the conference, and he was a good friend of George Housner. Muto planned the Second World Conference on Earthquake Engineering. Professor Kiyoshi Muto, in consultation with Professor Housner,

proposed to call the 1960 world conference the "Second" and inaugurate an ongoing series, and to establish the International Association of Earthquake Engineering to keep the series going. I later met Professor Muto at Berkeley on a world tour he was making, accompanied by half a dozen of his assistants, to become acquainted with the available earthquake engineering research facilities.

Eduardo Catalano, Eduardo Torroja, Felix Candela, and Pier Luigi Nervi

Bertero: There was a group at MIT that consisted of Stephen Crandall, Francis Hildebrand, and Eric Reissner in the mechanical engineering department. Reissner taught the theory of shells, a big topic at that time.

Regarding that topic of shells, I would like to mention an architect from Argentina who was at MIT at that time, Eduardo Catalano, and also three other experts who gave some lectures there but were not MIT professors. Catalano is famous for the hyperbolic paraboloid house of his that he designed in Raleigh, North Carolina when he was on the architecture faculty at the university there before coming to MIT. He came to MIT in 1956 and was on the architecture department faculty there for twenty years.

One day at MIT I received a telephone call from Professor John Wilbur. He said that Dr. Eduardo Torroja from Spain would be visiting. He asked me to assist him while he was here. I already admired him when I was in Argentina. Thus, it was for me a great pleasure to assist him, and to become a good friend of his. One of the finest honors that I have received is the first Eduardo Torroja medal, after Torroja passed away.[17] It is given by the Torroja Institute for Construction Science in Madrid. He was an amazing engineer-architect.

Then, Felix Candela[18] was to come to MIT. So again, they said, "Vitelmo, you know Spanish, you will assist him." Thus, again, this was a real pleasure for me to do so. Later Pier Luigi Nervi came to MIT and I discussed with him reinforced concrete structures.[19]

Reitherman: That's really something, to have been personally involved with Torroja, Candela, and Nervi. What was the very first step Candela would take in designing his adventuresome structures? Was it numerical, to put down some starting calculations? Was it drawings? Scale models?

Bertero: It wasn't numerical. It was more drawing and intuition, bringing out his theoretical background, with a feeling for his work. He had a tremendous influence in Spain and in Latin America. He had structural intuition.

17. Eduardo Torroja y Miret (1899-1961) was a Spanish engineer who specialized in the design of dams, bridges, aqueducts, and large-span building structures. In English, his structural design philosophy is described in *Eduardo Torroja y Miret, Philosophy of Structures*. University of California Press, Los Angeles, 1958.

18. Felix Candela (1910-1997) was a Spanish structural engineer who emigrated to Mexico and designed long-span thin-shell reinforced concrete structures.

19. Pier Luigi Nervi (1891-1979) was an Italian engineer and construction contractor who designed innovative long-span concrete roof structures that combined efficiency and beauty.

Reitherman: People say some similar things about Pier Luigi Nervi. He knew how to do structural analysis. He had first-hand knowledge of the contracting business and the construction process. He had a strong aesthetic talent. Based on all his experience, his first design step was intuitive, rather than analytical, only using analysis once the design was formulated in his mind. One story I heard from Joe Esherick when he was on the architecture faculty at Berkeley was about a visit Nervi made to Berkeley, when he gave a lecture for the architecture students. He talked about structural intuition, by which he meant the synthesis of his years of analysis, design, and construction experience. A young Berkeley architecture student made some comments about how wonderful design intuition was. Nervi quietly listened and then said, "I was referring to *my* structural intuition."

Bertero: Nervi was the same kind of creative designer as Torroja. In knowledge, perhaps, Eduardo Torroja had more knowledge about structural engineering.

Reitherman: Did you ever work on the design of any shells?

Bertero: I worked on the design of one, doing testing. In the end, I decided that I would not advise shells as the solution for earthquake engineering, and I did not want to pursue that field. The hyperbolic paraboloid has problems for earthquake applications. The structure is so thin that any mistake in construction can come out during the earthquake. They cover so much area with only a few supports.

Reitherman: How did you do a gravity load test on a shell? With distributed weights all over it?

Bertero: Yes, and with instruments all over it also. I remember I gave a paper on the tests at an international conference, and after I was done one of the experts in the audience said, "Professor Bertero, I believe that you have reinforced your shell with instrumentation." At MIT, I worked on the instrumentation of the concrete shell structure, Kresge Auditorium. But about half the instruments we installed were ruined when the workers stepped on them or damaged them.

Consulting Firms Founded by MIT Professors

Bertero: I have mentioned some professors at MIT in that time who became well known consulting engineers. It's really a rather impressive list. Robert Hansen, Myle Holley, Jr., and John Biggs formed Hansen, Holley, and Biggs. Howard Simpson, Werner Gumpertz, and Frank Heger formed Simpson, Gumpertz, and Heger. Jack Benjamin formed Jack Benjamin Associates.

There was also a young assistant professor, and this assistant professor, William J. LeMessurier, became very well known for his consulting structural engineering firm, LeMessurier Consultants, still located in Cambridge, Massachusetts. We called him Bill. I think he was from Harvard before he was at MIT. When I knew him, it was several years before he started his firm. He helped me a lot during my first research work at MIT, an experimental study of the behavior, or performance, of reinforced concrete beams using high-strength steel and concrete. This was a project under the supervision of Professor Holley.

Chapter 6

Earthquake Engineering and Other Disciplines

...that was the only way—to learn from other engineering disciplines about plasticity, dynamics, and probability.

Aeronautical Engineering

Bertero: Professor Charles Norris recommended that I listen to the lectures of Professor Raymond Bisplinghoff. I came to MIT to learn about materials, inelastic behavior, dynamics, so I could work on earthquake engineering. Professor Bisplinghoff was a professor of aeronautical engineering. He wrote a book on aeroelasticity.[20] He dealt with some important topics as applied to aircraft struc-

20.   Raymond Bisplinghoff, Holt Ashley, Robert L. Halfman, James W. Mar, and Theordore H. H. Pian, *Aeroelasticity*. Addison-Wesley, Reading, Massachussetts, 1955. Bisplinghoff left MIT for a time to work for NASA, returned to be the dean of the school of engineering at MIT, then was Deputy Director of the National Science Foundation. From that position he later became Chancellor of the University of Missouri at Rolla.

tures, not civil engineering structures, but he taught the principles very well.

It is true that generally aircraft are designed to remain elastic. Even a little bit of inelastic behavior, with so many cycles of loading, would lead to fatigue failures. But Professor Bisplinghoff taught about aeroplasticity as well as aeroelasticity. Aeronautical design was a university subject long before the scientific basis of earthquake engineering was established. The structures are of course quite different, but the principles are related. There was a greater need to be sophisticated in the design of airplane structures than buildings—to reduce the weight, and because the airplane always works in a dynamic environment.

That was an advantage of MIT. You could learn from good teachers in many different areas. Because earthquake engineering was not yet a university subject, that was the only way—to learn from other engineering disciplines about plasticity, dynamics, and probability. At MIT, I learned about dynamics, about inelastic behavior, and probability, but as I said, never took an earthquake engineering course.

Reitherman: Another example of how nonseismic engineering has influenced the earthquake engineering field, and one which is also MIT-related because it involves Ray Clough, who got his doctorate there, was when Clough co-invented the Finite Element Method, beginning work on that analysis method in the summer of 1952 for Boeing Aircraft Company. The problem was how to analyze jet wings of a delta shape. Of course, Professor Clough went on to become eminent in the earthquake engineering field, but that was later. Joseph Penzien worked for Convair on the structural

effects of blast on the B-36 airplane, and previously worked for Sandia Laboratory when it was doing experiments on blast effects on buildings. That was also well before Penzien became involved with the subject of earthquakes. Joe, like you, did his doctoral work at MIT on blast engineering.

Bertero: When we talk about the University of California at Berkeley, we can discuss the earthquake engineering that the faculty members there did. If we consider what happened within the U.C. Berkeley Department of Civil Engineering with respect to earthquake engineering—it became a leader in the field—it is remarkable that none of the original faculty who built up that discipline had been educated in that specific area. They had all been educated concerning topics that were related, and they were ready to tackle the problem of developing scientific solutions to the earthquake problem, to develop what is today *ingeniería sísmica moderna*, modern seismic engineering.

I have already mentioned some of the important people in the Mechanical Engineering Department at MIT, and there were mechanical engineers at other places doing work on dynamics. At MIT, there were Stephen Crandall and Jaapie, or Jacob, den Hartog. Stephen Timoshenko was at the University of Michigan and later Stanford. He is another example of a mechanical engineer whose knowledge of materials and dynamics later helped structural engineers with the earthquake engineering problem. Lydik Jacobsen at Stanford was a mechanical engineer who specifically applied his expertise to the earthquake problem.

Blast Engineering, Dynamics, and Inelasticity

Bertero: I have said how important the funding for blast engineering by the U.S. Army was, because there was no funding yet for earthquake engineering. A big engineering problem of the 1950s was the design of underground nuclear shelters, funded by the civil defense agency of the U.S. I explained that this was the purpose of the blast loading testing machine on which I worked in my doctoral studies. Generally, for ordinary loading, like gravity every day, you do not think that any damage is acceptable. For hazards that may be very extreme or may not even occur, however, you typically accept the chance of some damage. For the case of blast shelters in the event of war, when you say that some damage is acceptable, in structural engineering terms you are thinking about inelastic behavior. The whole purpose of a blast shelter was for people to go there for a while and maybe the structure would have to resist tremendous loads. If I have to design a blast shelter to withstand such big loads without any damage, I do not know what I could do. Inelastic response and ductile characteristics of the materials make it feasible.

Reitherman: Would you explain the connection between the engineering for that kind of blast loading compared to earthquake loading? What could you later apply to earthquake engineering that you learned from those defense-related studies?

Bertero: The Finite Element Method was new then. There was at MIT in 1956 a young scholar from England named Chris Calladine, a theoretical researcher who worked analyti-cally in the design of structures using computers. MIT had recently set up the Computation Center, and Calladine was working with Professor Robert Hansen on numerical analysis of reinforced concrete shear walls subjected to blast loading. So we had to compare with experimental results. At that time, the Finite Element Method was still being developed by Ray Clough, and was later to become widespread as computer capabilities developed.

You can analytically replace a shear wall with a number of little trusses. Calladine did the theoretical work at MIT to compare that with the experimental results we were getting. Joseph Antebi, a young doctoral student then, who now is a vice-president of Simpson, Gumpertz, and Heger, went on after my work at MIT to do his doctoral degree at MIT on columns and other structural elements.

The civil engineering research being funded by the government at that time at MIT was on underground shelters. It was necessary to go underground for shelters to avoid the dynamic pressure through the air, like a terrible wind, and the radiation. You could not produce practical designs to resist those effects above ground if you were near the zero point where they would detonate a nuclear weapon. For me, these large loads and the design of structures to resist them was an important line of research that provided information for earthquake engineering, though the two kinds of design are different. The loading from the blast wave is just "boom!" That's it, one huge pulse. Close to a nuclear detonation you are dealing with tremendous pressures that require consideration of inelastic behavior of the structure. The blast wave propagated through the atmosphere into the ground near the detonation, then traveled through the ground to hit the underground shelter as a soil pressure all around it. The blast wave through the atmosphere also traveled to the site and hit the soil directly from above. You had these tremendous pressures on the walls of the shelter, so we were testing shear walls to find out how they behave under the lateral forces they could be subjected to. The first two summers I was on the faculty at Berkeley, I had to give a short course on civil defense, on the design of shelters. I remember very well what one of the students said on their course evaluation form: "Professor Bertero is a good instructor, but at the beginning I found it hard to understand his combination of Argentine Spanish and California English."

Reitherman: How is the very brief pulse of blast wave loading through the soil different than earthquake loading?

Bertero: The only big difference is the duration of the pulse and the number of the pulses. In the case of just one blast, we are talking about milliseconds of duration with the blast loading.

At that time, the basic earthquake design approach, which came from Japan, or from Italy, or from both, was to apply a seismic coefficient and then compute a base shear and distribute it up the height of the structure as static lateral loads intended to be equivalent in an overall design sense—considering the allowable stresses that were used, with everything idealized as existing in the elastic range of loads, and so on—to the effects of the actual dynamic loads.

I started to look at earthquake ground motions as a series of pulses.

In 1971, during the San Fernando earthquake in Los Angeles, important ground motions and structural responses were recorded. But one of them, recorded on the abutment to a dam, had a peculiar feature. Are you familiar with the dam there?

Reitherman: Pacoima Dam?

Bertero: Yes. That strong motion record had a very strong and long duration pulse. I immediately conferred with Professor Bolt, Bruce Bolt. Many people said there was something wrong with the way the instrument recorded this data, and that the record was invalid. But Professor Bolt found an explanation for such a pulse. Professor Ray Clough also worked on the justification of the data so that people would take it seriously. Yes, there was a strong pulse in that record.

I became convinced that you could study the effects of earthquake ground motions by considering them as a series of pulses, and that relates to your question about what we learned from the blast engineering research after World War II.

For that reason, I said that I did not get an education in earthquake engineering at the Massachusetts Institute of Technology, but what I studied was so closely related to it that it was very relevant. Two different phenomena: one was a single massive pulse, starting in the atmosphere or at one point on the ground, caused by blast. The other was an earthquake, a series of pulses sent through the ground. In both cases, the only practical design approach was to rely on the ductility of a structure.

Relationship of Inelasticity to Dynamic Response

Bertero: The problem with a series of pulses in an earthquake, rather than a single pulse in a blast, is that if all the pulses have the same dynamic characteristics and the structure remains linearly elastic, the repetition of these pulses can lead to the phenomenon of resonance, and to collapse. However, if the material and the structure are ductile—that is, can undergo inelastic (or plastic) deformations—the period of vibration of the structure will change as soon as some critical regions of the structure start to yield, and the resonance will disappear if it is a system of just one degree of freedom. In mechanical engineering, typically a machine has to stay completely elastic, and then any tendency to resonance builds up tremendous forces. But with structures responding to earthquakes, as soon as a member or connection undergoes inelastic (plastic) behavior, that lengthens the period of vibration of the structure and tends to move the structure out of the resonance. I said earlier that in my first class at MIT with Professor Holley, he told us that ductility was the blessing for the structural designer.

Therefore it became clear to me that in earthquake engineering, the first insight was that the structure needed inelasticity because the forces were so high it needed that reserve capacity. But as ground motions and structural response were better understood later on, engineers understood that inelasticity not only helped the structure endure large earthquake loads, it also affected how large the loads would be.

Generally, it would not be economically acceptable to design structures to remain elastic, without any damage at all, under the probable maximum earthquake ground motion that can occur. The earthquake engineering field began to realize the desirability of inelastic deformation and how to manage it. The understanding of inelasticity gave the designer a way to control the forces, and thus the strength, that the structure needs.

Reitherman: When did this aspect of ductility, that it changes the period of the structure and keeps it from resonating, become commonplace knowledge in the earthquake engineering field?

Bertero: Most earthquake engineering thinking in the early days was working only on the question of elastic behavior. In Japan, for example, the idea of keeping the structure rigid and maintaining its period was a design principle. It was in the 1950s that structural engineers started looking at the fact that if the material was ductile, the stiffness of the whole structure changed, and if the stiffness changed, you have a completely different problem.

Reitherman: At the zero period point on the response spectrum, for a one-degree-of-freedom system that is essentially rigid, the response is the same as the peak ground acceleration. The typical elastic spectra curve shows the response rapidly increasing from the zero period to a plateau beginning in the range of 0.2 second to 0.5 second. Suppose you designed a building to have a very short period, say 0.1 second, because it was a short and relatively light building with extensive walls, giving it high stiffness compared to its mass. Picture a small building like a one- or two-story house, with every wall carefully arranged and also made into an unusually strong shear wall. If you could give it enough strength to keep it from going inelastic, it would keep its short period, and therefore its response would not climb up the response curve. An infinitely rigid structure by definition would do exactly what the ground did, not experience worse motion because of amplification. Would this be a valid seismic design approach for limited cases?

Bertero: Theoretically it may be possible, if the structure could remain completely elastic. But because of economic and social requirements, and for practical purposes, it will not be possible to design and construct an infinitely rigid building that will be habitable and that will remain infinitely rigid during severe earthquake ground motions. Any practical building will loosen up and start to climb up the elastic response curve. So for practical design, it is not only desirable, but necessary, to let the building undergo some inelastic (plastic) deformations—that is, provide the structure with sufficient physical ductility. These plastic deformations should be controlled because they are associated with damage to the structural materials that are at present commonly used. The physical ductility allows the designer to overcome the many uncertainties that he has in the design—uncertainty in the ground motion, uncertainty in his design approach to the structure, and uncertainties on how the building will be constructed and maintained.

We also need to say that here we are talking in generalities, but you have to look at the spectrum for your specific site. On the lakebed of Mexico City, with very deep and soft soil, if your building period lengthens from, say, one second or one and one half seconds toward the two-second range, its response would increase, because the response curve has a peak at about that period for that particular site. That is unusual, but it illustrates that you have to consider the spectrum for your site.

Reitherman: Climbing up the response curve from the zero period level to the peak, at half a second in this example, seems like an unstable equilibrium, like a ball on top of a dome. As soon as you displace the ball a little bit, it is unstable and will roll faster and faster down the side of the dome. Things get worse quickly.

The EERI Oral History Series has a very wide readership. Perhaps for readers without an engineering or physics background you could explain the period of vibration a little more.

Bertero: The mass of the building doesn't change during the earthquake, so when the stiffness reduces because of inelastic behavior, the mass/stiffness ratio increases and the period lengthens. The period increases as the square root of the decrease in the stiffness. If the stiffness is decreased by a factor of four, the period increases by a factor of the square root of four, or two. If the total building system can deform inelastically, it will not stay at its original period, but will quickly move to the right on the x-axis on the response spectrum graph—the period increases. If it is at the peak—again for discussion we will say this is one-half second, though this is not always the case—it will soon slide off to the right side of the peak, down the response curve, lessening its response as its period lengthens. This is in the context of a system that has a single degree of freedom.

This was one of the insights that marked the time in the 1950s and 60s when earthquake engineering started becoming scientific—when people started looking at the real behavior of a total structural system as it went through an earthquake, second by second. The equivalent static lateral force method, however valuable it has been, tends to make the designer think of one lateral load, the one they calculate, the base shear that is then distributed up the height of the building. The earthquake is a series of lateral loads on the structure, spread out over ten, twenty, thirty seconds, or more than a minute in the case of large earthquakes like the 1960 Chile or 1964 Alaska earthquakes. The designers should try to see their building responding to that continuing ground motion and realize the building changes during that time. Design engineers typically only see one set of lateral loads in their calculations—they see only one building with the same constant mechanical characteristics. That's not the reality.

To keep the design process simple, practicing engineers would like to still design on an elastic basis, the region where stress is proportional to strain, because on the basis of that principle many calculations are greatly simplified. However, there are significant problems with reducing actual expected loads, and/or deformations, which will make the building behave inelastically with various code-specified factors, so the engineer can use elastic analysis. This is the problem of the R Factor, which we can discuss further.

When I was in Argentina in the 1940s, the education was all about elastic behavior. For this reason I came to the United States to learn about inelastic behavior because that is what is going to happen to a structure anywhere in the world with severe earthquake ground motions. We have still not educated the professionals involved in seismically resistant design and construction sufficiently to think in terms of inelastic response in case of significant earthquake ground motions.

Ultimate Load Design and Plastic Behavior

Bertero: In the United States before World War II, Charles S. Whitney and J. A. Van den Broek introduced the ideas of ultimate load design and plastic behavior You have to understand plastic behavior to develop ultimate load design methods. Whitney worked with concrete.[21] Van den Broek focused on steel.[22] Their work was initiated prior to World War II. Whitney and Van den Broek should get credit for their pioneering work. Then, it was two Englishmen who really made the practical applications of the ideas under the pressure of World War II.

---

21.   Charles S. Whitney, "Plastic Theory of Reinforced Concrete Design," *Proceedings of the American Society of Civil Engineers*. December 1940; see discussion of the paper in *Transactions of the American Society of Civil Engineers*, Vol. 107, 1942, pp. 251-326.

22.   J. A. Van den Broek, "Theory of Limit Design," *Transactions of the American Society of Civil Engineers*. Vol. 105, 1940. Also a book of the same title published by John Wiley and Sons, New York, 1948, in which he connects the concepts of limit design and plastic behavior: "The theory of limit design presupposes ductile or semiductile stress distribution. In it, emphasis is shifted from permissible safe stresses to permissible safe deformations." (p. v)

In England, this design approach was not developed because of the earthquake hazard, but because of the urgent need in World War II to use materials efficiently in constructing factories, bridges, and so on. In our earthquake engineering field, we adapted the concepts much later. When you consider plasticity, you can design much more economically, while still designing safely. Cambridge University and Imperial College had some very good researchers working on this problem.

Reitherman: So ultimate load or strength design, or limit state design, came from a need to be more economical—originally just for gravity loads, not because there was a concern over danger or a demand for greater safety?

Bertero: Yes, it was a push for greater efficiency, especially to build the large number of factories, bridges, and so on that were the backbone of the country's productivity. But it was not just for gravity loads alone, but also for other forces such as those induced by winds, changes in temperature, bomb explosions, and so on. Even though economy may be the original motive, you also attain greater reliability in estimating the safety of the structure by means of this type of design and analysis.

The English were working hard to make their structures more efficient, through plastic design and analysis. There were the two Bakers. The Baker who worked on steel, J. F. Baker, was at Cambridge. That came first. The Baker who worked on reinforced concrete, A. L. L. Baker, was at Imperial College.[23] At Cambridge Uni-

versity, under the direction of Professor J. F. Baker, there was a group of younger researchers such as Horne and Heyman, who conducted experimental work supporting the theory of plastic analysis. Baker, Horne, and Heyman in 1956 published an important book on steel design.[24] Other English researchers, such as Neal and Symonds, did important research on plastic analysis.

Much later, in 1965, I was in Europe and was amazed that there was a great difference in which countries were accepting limit state design. The Germans and Swiss wanted no part of it. They were very much opposed. Already in 1935, the very well known professors at the *Eidgenössische Technische Hochschule* (the Swiss Federal Institute of Technology) in Zurich— F. Stüsi and C. F. Kollbruner—published a paper in the publication *Bautecnik*, which clearly showed that the use of limit state design based on just rigid-perfectly-plastic models, as had been proposed earlier, can be misleading about the estimation of the ultimate load.

In 1965, I remember being at one conference to present a paper. I explained that the history of forces and deformations that occurred over

---

23.   J. F. Baker was a professor at Cambridge University. One of his writings aimed at the design engineer was: J. F. Baker, "The Design

of Steel Frames," *The Structural Engineer*, Vol. 27, October 1949. A. L. L. Baker was a professor at Imperial College of Science and Technology. A summary of his work on ultimate-load theory is in A. L. L. Baker, *The Ultimate-Load Theory Applied to the Design of Reinforced & Prestressed Concrete Frames*. Concrete Publications Limited, London, 1956.

24.   J. F. Baker, M. R. Horne, and J. Heyman, *The Steel Skeleton, Plastic Behaviour and Design*. Vol. 2. Cambridge University Press, Cambridge, UK, 1956.

the seconds of duration of an earthquake was important for predicting how the structure would behave. I did not say that plastic design was bad—I was an advocate of that method. I was just saying that if you had to control deformation, you had to be careful defining the dynamic loads. Then Professor H. Rüsch, a great German teacher and designer of reinforced concrete structures, immediately said, "You see? Professor Bertero clearly shows why he is against limit state design." Oh my! I was just explaining that you had to be careful with that method, like any method, and pointed out some particular considerations. I was actually in favor of the limit state thinking, because it was more realistic.

In 1964, I was a contributing author to an American Society of Civil Engineers-American Concrete Institute (ASCE-ACI) publication that covered the papers and discussions presented in an International Symposium in which plastic analysis and design of reinforced concrete were discussed.[25] Professor Herbert Sawyer was the leader of that effort. ACI publicized the symposium and the ASCE staff did the editorial work. It took a number of years for that approach to be applied in the

---

25.   Herbert A. Sawyer, Jr., ed., *Proceedings of the International Symposium on Flexural Mechanics of Reinforced Concrete*. November 10-12, 1964, American Society of Civil Engineers.

Chapter 7

The University of California at Berkeley

I think the Berkeley civil engineering faculty became so strong partly because the individuals worked so well together as a group.

Bertero: After June 1957, when I received my doctorate, as I was completing the year I had promised to spend in the laboratory at MIT, assisting with the continuing blast engineering research, I had to consider my future. The secretary of Professor Charles Norris had arranged the details for me to get a permanent residency visa, or green card. I thought it would be necessary to return first to Argentina and to apply for it there, but all I had to do was make a brief trip out of the country to the U.S. embassy in Montreal. An hour after I got there, I had my permanent residency card, so it was much easier than I had thought.

Professor Norris said that I could have an assistant professor position at MIT, but he was not sure that the direction of the Civil Engineering Department would be parallel with my interest in structural design.

In Argentina, I had an offer from Bahia Blanca University, but the situation of the universities in Argentina then, in the end of the 1950s, was worse than a decade before, when I was at the university in Rosario. There was an effort at that time in Argentina to establish private universities, rather than have them all run by the government. There was a great deal of resistance to this, because some people considered the development of such universities a threat. They might have a religious orientation and support. They might be connected with a foreign nation like the United States. Students at the government's national universities had many strikes against the development of private universities, because they did not want the competition or some other reason. I could also go back to my old university in Rosario, but the government had replaced the university administrators with their own people and the situation was not a good one.

I talked over the situation with my family and they thought it would be better that we stay in the United States.

Visiting Cooper Union

Reitherman: What universities did you consider?

Bertero: A student of Professor Norris had become the dean of Cooper Union. Norris suggested I meet with him because he knew they were going to hire an engineering faculty member at the end of the year, when I would be done at MIT. I had not heard of Cooper Union, and Professor Norris said that it was only a small college, but it was high quality

and provided many undergraduate students who went on to get their doctoral degrees at Columbia University.

So, an appointment was arranged and I traveled to New York City, to Greenwich Village where Cooper Union is located. I was shocked. I had an 8:30 am appointment with the dean. The sidewalk in front of the building entrance was littered with bottles and trash. A guard came to the door, asked me what I wanted, and I said I had an appointment with the dean. He took my letter, had me wait till he verified I had an appointment, and then he let me in.

In the meeting with the dean, which lasted two hours, I thought they had a very good engineering program. Then he gave me a tour. We went into the library, and I was shocked again. There were adults from off the street sitting there, knitting to mend their clothes, spending the day there to enjoy the air conditioning inside. As a requirement of the will that gave Cooper Union its money to become established, the library had to remain open to the public. In the winter, there were also people from the street in the library because then it was cold outside and it was heated in the library.

Reitherman: So, in other words, the college's library was, in effect, what we would call a homeless shelter today?

Bertero: Yes, it was an unusual situation. They made me a good offer, but I said I needed to know where I could rent a house for my family. The dean said he did not advise me to have my home, with my wife and four children, near Cooper Union. I went around the neighborhood and looked and he was right. Today Greenwich may be a fashionable district with a Bohemian reputation, but then it looked to me to be inadequate for my family.

Visiting Chicago

Bertero: Back at MIT, there were two other faculty openings they told me about. One was at a research institute in Chicago, which I think was operated by the Illinois Institute of Technology and later was combined with the overall university.

I knew about the research institute in Chicago because a fellow Argentine, Augusto Durelli, was there. After Durelli received his undergraduate degree from the University of Buenos Aires and a doctorate from the Sorbonne, he spent some time at MIT on a fellowship. This all happened before I was at MIT. He went back to Argentina, but he left right after World War II. He was an outspoken man, and he wrote an article that in English is called "The Colonel's Backpack," a way to refer to Juan Peron and criticize him without referring to him by name, but everyone knew who it was. I had some correspondence with Professor Durelli when I was at MIT. He became famous in the field of photoelasticity, and in those days, that was a very popular topic.

This was in December of 1957. I traveled to Chicago, and the research institute made arrangements for me to stay in a nice hotel. But when I took the elevated train and got closer and closer to the research institute's site, the neighborhood got worse and worse.

Reitherman: This sounds like your experience visiting Cooper Union.

Bertero: Exactly. I was a little scared going there. Professor Durelli told me that I could not find a good place to live near there with my family, it would have to be somewhere else. He offered me a good job of teaching and research.

Visiting the University of Illinois at Urbana-Champaign

Bertero: I have mentioned how Nathan Newmark worked with Hansen, Holley, and Biggs. So it was recommended to me to visit the University of Illinois at Urbana-Champaign to see about a faculty job there.

The day after I visited Professor Durelli in Chicago I flew to Urbana-Champaign. It was a nice day, and the next morning I went to see Professor Newmark. I also met with other faculty: Chester Siess, William Hall, and Norby Nielsen. I also met for the first time Mete Sozen, who then was a student who was just completing his PhD. I think that Sozen did a large amount of the work for the well-known book by Blume, Newmark, and Corning on seismic design of reinforced concrete.[26] I think Professor Siess also did work on that book.

They gave me a nice tour of the laboratory, which was a very fine laboratory. I was impressed. At that time, their main line of research was funded by agencies like the U.S. Army on shelters, similar to the topic of the

---

26. John Blume, Nathan Newmark, Leo Corning, *Design of Multistory Reinforced Concrete Buildings for Earthquake Motions.* Portland Cement Association, Chicago, Illinois (PCA is now located in Skokie, IL), 1961. Re-published in a 1992 edition by PCA.

research I had worked on at MIT. So they were interested in that.

Reitherman: I think the shake table at the University of Illinois was installed in 1967 or 1968, so it wouldn't have been in the lab yet.

Bertero: No, there was no shake table. They had a large testing machine for loading full-size specimens of reinforced concrete. They had very good equipment.

They talked to me about taking a position at the university. I said I would think about it. That night, it started snowing. My flight the next morning was cancelled. The train was also cancelled. I had to stay two days in Urbana-Champaign. I would not have had a problem of getting a good house for my family in that town, but getting stuck there for two days was not a good experience. It discouraged me.

Back at MIT, Professor Norris said I had time to think it over.

Visiting Lehigh University

Bertero: There was a Russian émigré at Lehigh named Alexis Ostapenko, who had been at MIT, where he got his doctorate about a year before me. He had been a prisoner in Germany during World War II, was in the part of Germany where the American Army came in at the end of the war, and then he came to the U.S.

Lehigh had the best group of people working on steel. One of the professors, Bruno Thurleman, was no longer there, having returned to Switzerland. The professors who were leaders in the plastic design of steel who were at Lehigh when I visited were G. Driscoll,

John W. Fisher, and Lynn Beedle, who was a student of Egor Popov. Ted Galambos was just getting his doctorate degree there. They were very nice to me and gave me information about their research that was very helpful to me. When I started to teach, when I joined the faculty at the University of California at Berkeley, I found that information very helpful.

Hired by the University of California at Berkeley

Bertero: Now I mention Professor Charles Norris again. He told me in 1958 that Professor Egor Popov from Berkeley was making a trip across the U.S. to look for good people to hire. Professor Popov was looking for someone to teach structural design, and as of then, Berkeley was already involved in the earthquake engineering field a little bit. You recall that in 1956 the first of the World Conferences in Earthquake Engineering had been held at the University of California at Berkeley. Professor Norris knew I wanted to specialize in earthquake engineering. Professor Norris knew of Egor Popov, Alexander Scordelis, Ray Clough, and Joseph Penzien because they had been at MIT, and he recommended that it would be good to work with them.

Professor Norris thought it would be a good opportunity for me and arranged for me to meet with Professor Popov. It was March of 1958, often spring weather in Massachusetts, but it was a very cold day. After we met, Professor Norris asked me to take Professor Popov to the faculty club for lunch. It was about four blocks away, and it was snowing. The Charles River was still frozen. Professor Popov commented on how cold it was. I asked what the weather was like in Berkeley, California. Professor Popov said, "If there is a paradise on earth, it is Berkeley. The weather in California is perfect." I laughed and said, "Professor Popov, I know you are trying to encourage me," but he said he was not exaggerating.

We had lunch, he toured the laboratory, and at the end of the day he said he would recommend to his fellow faculty that I be offered a position at Berkeley. In less than two weeks, I received an offer. I would start right away as a lecturer for two years. I would not have tenure, and if it worked out I would then get a tenured position. Later I learned from Professor Popov that Professor Norris had told him that the best candidate for the Berkeley job that he could recommend—who also had the worst English!—was me. Professor Popov and his colleagues knew how I would fit in the program being developed at Berkeley.

How things have changed! Today, it's all a lot of correspondence, documentation, presentations, and committee meetings to consider candidates. I know all these problems because I was the chair of a committee evaluating new professor candidates. We spent more time justifying why we were not selecting people, especially depending on background and affirmative action criteria, than we did seeking good people that would be the right addition to our particular program.

Tenure meant the associate professor rank, the usual U.S. system. It was the same at MIT in principle, but actually it was a little different. At MIT they had lots of assistant professors who were there for five, maybe six years. Some were doing mostly research, only a little

teaching. Only a few would get tenure position offers, and that was not a good situation. It is better to tell assistant professors within a couple of years whether they will have a future on the faculty.

You may know that Professor Popov was the one who proposed and fought for the development of the doctoral program in civil engineering at Berkeley. Mihran, or Mike, Agbabian was the first structural engineering doctoral student at Berkeley, working with Professor Popov. Professor Popov built up the Berkeley educational program in graduate studies and the research side to it, and he was a key person in hiring others like me. Later, in 1977, it was an honor for me to be one of the five members of the Symposium Committee honoring Professor Popov. The papers presented at the Symposium were published in a volume edited by Professor Karl S. Pister, who was the chair of the Symposium Committee.[27]

There were five of us in the Structural Engineering and Structural Mechanics (SESM) division in the Civil Engineering Department at Berkeley who had come from MIT. Egor Popov was hired in 1946, Alexander Scordelis and Ray Clough in 1949, Joe Penzien in 1953, and then I was appointed a lecturer in 1958 and then associate professor in 1960.

Professor Popov was educated at Berkeley as an undergraduate student, then received a master's at MIT, then began his doctorate at Caltech with R. R. Martel and finished it at

---

27. Karl Pister, ed., *Structural Engineering and Structural Mechanics: A volume honoring Egor P. Popov.* Prentice-Hall, Englewood Cliffs, New Jersey, 1982.

Stanford, working with Stephen Timoshenko and Lydik Jacobsen. So Professor Popov had first-hand experience with how engineering was done at Berkeley, MIT, Caltech, and Stanford. I think he felt that at that time MIT had the strongest program in educating civil engineers to be members of the structural engineering profession.

Reitherman: In the Popov oral history,[28] we see that he says that when he had John Wilbur as his master's thesis advisor at MIT, he realized that his undergraduate civil engineering education at Berkeley was not as strong as it should have been.

Did Popov and his colleagues search for someone to just teach structural design and improve the department's capabilities in that area, or also specifically to do earthquake engineering teaching and research?

Bertero: Although they wanted someone to teach structural design, the chair of the SESM division and the Department of Civil Engineering told me when I arrived at Berkeley in 1958 they were also interested in someone who would conduct research in the field of the earthquake-resistant design of structures.

Early Years of EERI

Bertero: Back in 1958, I think the Berkeley Civil Engineering Department was unusual in wanting to hire someone to do earthquake engineering research and teaching. That was very unusual then. You know, Ray Clough already

28. *Connections: The EERI Oral History Series—Egor Popov*, Stanley Scott, interviewer. Earthquake Engineering Research Institute, 2002, p. 37.

was very much involved in the earthquake field, even if there wasn't research grant money for that field yet. He was a key organizer of the World Conference in 1956. He was very much involved with the Earthquake Engineering Research Institute then.

Back in those days, the first decade of EERI's existence, there were very few members, and they were all invited to join, you didn't apply.

Reitherman: The data point I know is that when I joined in 1978, EERI already had about 700 members, and although you needed to provide three references and wait six months or so for your application to be evaluated, it was no longer an exclusive, small group of top experts who were invited to join.

Bertero: And it kept getting bigger and easier to join. George Housner was a key advocate of enlarging the organization. John Blume thought it should stay small, so it could function like one committee.

Reitherman: I've talked with Joe Penzien about that. Joe was the Chair of the Bylaws Committee of EERI when in 1973 they were changed to allow people to apply for membership, rather than being invited to be a member. Apparently, he and George pushed to open up the membership, but there was some reluctance to change things. In retrospect, it seems that if EERI had not opened up and expanded, there would have had to be some other umbrella organization to play that role. It's interesting that from the start, even when it was the size of a large committee, there were both professors and practicing engineers in the organization.

Bertero: Ray Clough was one of the key academic people in EERI back then, along with Professor Housner of course. EERI needed the connection to what was going on in the universities, but there was some tension between the university role and the practitioner role. I think that John Blume and Henry Degenkolb in particular were not happy sometimes with the academic people. Professors Karl Pister and Boris Bresler got into some spirited discussions with them at EERI meetings, I recall.

In 1969 I was accepted as a member of EERI. In other words, I was invited to be a member. There were maybe fifty members, and we could meet in one room easily. There were committees that also met. There were very few academic people. Nathan Newmark, Anestis Veletsos, and Bill Hall were there from the University of Illinois; I think Mete Sozen, also from Illinois, started getting involved in EERI activities a year or so after that. There were several faculty from Berkeley, and half a dozen from southern California, people like George Housner, Martin Duke, Donald Hudson, Paul Jennings.

Arriving at Berkeley

Bertero: In July of 1958, I moved to Berkeley with my family in a cross-country drive. That was when I realized why the United States was such a rich country—so many crops, so many cities. It was a nice trip by car to see the country.

Professor Popov took me to meet the chair of the division within the Department of Civil Engineering, which was called then Structural Engineering and Structural Mechanics, SESM.

The chair of SESM was T. Y. Lin, who really surprised me. He said, "We know that you are a good teacher, so the only thing you have to work really hard on is doing good research." Good teacher! The most teaching I had done was back in Argentina some years before. I don't know why the Berkeley faculty trusted that I could become a good teacher, except for my recommendations from MIT faculty.

I was able to rent a four-bedroom house in a good neighborhood, and which was near the campus.

I was assigned to share the office of Professor Joe Penzien. This is before the current Davis Hall was built. Professor Penzien was very nice to me. I was lucky that Joe was then working on some research on plastic design, so we had some very good discussions.

At lunchtime, often other civil engineering faculty would come there—the "brown bag" group who brought their lunches there to have lunch together. The usual group was Joe Penzien, Ray Clough, Jack Bouwkamp, and Alex Scordelis. Two civil engineers outside the structural area were also usually there: Professor Carl Monismith, a transportation engineer, and Francis Moffitt, who taught surveying. I remember when I met Professor Scordelis, who taught structural analysis, that he asked me what textbook was used for that course in Argentina. I said it was the Fife and Wilbur book that I mentioned earlier. He asked me several questions about what I thought of that book. I said that it was a good book, but it was lacking in two areas. It treated structures as a combination of plane elements, not as three-dimensional systems, and it was limited to elastic analysis. Professor Scordelis was later to do

Table 1. *Graduate Students Supervised by Vitelmo Bertero*

PhD Degree Students

Atalay, M. B.

Anderson, J. C.

Aroni, S. (with B. Bresler)

Alussi, A. (with M. Polivka)

Axley, J.

Barez, S.

Bertero, R. D. (U. of Buenos Aires, with Alberto Puppo)

Charney, F.

Filippou, F. (with E. Popov)

Guevara-Lopez, T. L. (with H. Lagorio & H. Rittel)

Guh, T. H.

Hidalgo, P. (with R. Clough)

Kamil, H.

Klingner, R.

Krawinkler, H. (with E. Popov)

Kustu, O. (with J. Bouwkamp)

Lara-Montiel, O. (U. of British Columbia, with Carlos Ventura)

Lee, H.-S.

Ma, S. M.

Moazzami, S.

Moustafa, S. A.

Mahin, S.

Malik. L. E.

Miranda, E.

Ozselcuk, A.

Sasani, M. (with A. Der Kiureghian)

Sedarat, H.

Selna, L. (with B. Bresler)

Soleinami, D. (with E. Popov)

Teran, G. A.

Uang, C.-M.

Vallenas, J.

Vasquez, J. (with E. Popov)

Viwathanatepa, S. (with E. Popov)

Wang, T. Y. (with E. Popov)

Whittaker, A. S.

Zagajeski, S.

Master's Degree Students

Almant, R.

Alonso, L. J.

Bertero, R. D.

Brito, J.

Broken, S.

Calvi, M.

Chavez-Lopez, G.

Chandramouli, S.

Chowdhury, A. A.

Cova, A.

Cowell, A. D.

Felipa, C.

Fierro, E. (with E. Popov)

Forzan, B. (with E. Popov)

Galunic, G. (with E. Popov)

Gonzalez, G. (IISEE)*

Herrera, R.

Hollings, J.

Irragory-Montero, G. J.

Javier, A.

Javier, C.

Lara-Montiel, O.

Manrique, M. L.

McClure, G.

McGuire, R.

Megget, L. (IISEE)*

Meltzer Steinberg, H. (IISEE)*

Oñate, E.

Pereda, J.

Roha, C.

Sanches, E.

Saavedra, M. (IISEE)*

Sandoval, M. R.

Svojsik, M.

Sause, R.

Shahrooz, B.

Taheri, Ali

Thompson, C.

Uzcategui, I.

Verna, R. (IISEE)*

Villaverde, R. (IISEE)*

Wagner (later Phipps), M..

Zeris, C.A. (with S. Mahin)

* IISEE: International Institute of Seismology and Earthquake Engineering, Japan

some very good work in the 3D field, to model and analyze structures the way they really are. I knew some people at the Torroja Institute in Spain were extending the Hardy Cross frame analysis method to frames of towers loaded in two directions at the same time. Some of these things may seem obvious today, nonlinear inelastic analysis and three-dimensional analysis, but in the late 1950s, these were very new areas.

Then later I moved to an office above the big testing hall. That was the one that existed before the present Davis Hall testing laboratory was built.

Reitherman: Speaking of those days, Professor Bertero, I've heard Joe Penzien describe how he and others of your generation called you by the nickname Vic when you arrived at Berkeley, whereas the short form of your first name, Vitelmo, would be Vit, and I've also heard some people call you Vit.

Bertero: The middle name given to me by my parents was Victorio. Victorio in Spanish is the same as Vittorio in Italian, so whether I have the nickname Vic or Vit doesn't matter to me.

Teaching a Variety of Courses

Bertero: In the early years at Berkeley, each civil engineering faculty member would teach several different courses. When the instructor educates students in a variety of subjects, it is a good education for the instructor. So I taught statics, strength of materials, steel and concrete design, indeterminate structures, experimental methods, solid mechanics. When I started at Berkeley, there was no course on earthquake engineering, and I started that after I introduced the course on inelastic behavior of structures, or plastic design. The student had to take those other courses before taking the seismic design course, which was then more broadly called Structural Design for Dynamic Loads.

What I was used to in Argentina was a different situation. A professor would teach the exact same course every year, always the same course. With that system, the professor did not learn what the other professors were teaching, and what the students were learning in their other courses. At Berkeley, we could make sure the students had available to them the right sequence of courses to take, without duplications or gaps. So, when Ray Clough started to spend more time on the Finite Element Method, I became the instructor of the course he previously taught on experimental stress analysis, which Professor Howard Eberhart also had taught. Later on, when William Godden joined the faculty, he started to teach that class. Professor Eberhart had lost a leg in World War II, and he did research on how to design prosthetic devices. He told me that in addition to dealing with injuries like his, he was looking toward the day when the population would be much older. He knew that people would be living longer and longer, and yet even if medical techniques made the rest of the body last longer, the joints were basically a mechanical system that would wear out.

I also taught a civil defense course for engineering faculty from other universities around the country in the summertime, where the topic of the design of nuclear shelters was discussed. I was asked to do that for three or four years.

At one time, I taught solid mechanics. Also engineering materials, steel design, concrete design—teaching all those things was an education for me. Professor Boris Bresler was teaching reinforced concrete at that time, and I worked with him on that, teaching an advanced course. Bresler was very interested in plastic design. Professor Bresler also did research on steel structures, and authored a textbook with Jack Scalzi and T. Y. Lin on steel design.[29] Professor Bresler left the university to set up the Wiss, Janney, Elstner office in California, near Berkeley in Emeryville. Bresler then hired Sigmund Freeman, and the office has grown since then. I started to teach reinforced concrete more after Bresler left the faculty.

Later, there was a new advanced-level course I taught on the inelastic design of steel structures. When Jack Bouwkamp left to return to Europe, we had one vacancy on the faculty in steel design, and that was when I taught that course, in the 1980s.

One of the techniques I used with my graduate students doing research under my supervision was one that I learned from MIT. Each doctoral student had to give lectures or seminars to the other students several times. It was an effective technique.

Origin of the Earthquake Engineering Course at Berkeley

Bertero: In 1966, I prepared the material on a graduate course on inelastic behavior that I was given the opportunity to teach. The course

was given for the first time in 1967. It was originally called Inelastic Design and Analysis of Structures. Then in 1968, I started to teach a course devoted to earthquake engineering, though it was originally given a more generic title about design for severe dynamic loading.

This was just after Robert Wiegel organized a short course that turned into one of the early textbooks on earthquake engineering.[30] Most of the authors of the chapters in that book were Berkeley faculty, but there were people from Caltech also, like George Housner and Don Hudson, and practicing engineers like John Blume and Henry Degenkolb.

Reitherman: Quite a blue ribbon list of experts. Does the earthquake engineering course that you started in 1968 continue to the present at Berkeley? Is it the earthquake engineering course for master's students that Steve Mahin teaches today?

Bertero: Yes, though the course numbering and title have changed, and over the years the university went from the semester system to having three quarters in the regular academic portion of the year with a fourth quarter in the summer, and then it switched back to the current system of semesters. I taught it until I retired in 1991, and since then Professor Mahin has taught it.

Reitherman: It is commonplace today for a university to offer a master's level course in earthquake engineering. It's a striking contrast between eras. You developed and taught an

---

29. Boris Bresler, T. Y. Lin, Jack B. Scalzi, *Design of Steel Structures.* John Wiley and Sons, New York, 1960.

30. Robert L. Wiegel, editor, *Earthquake Engineering.* Prentice-Hall, Englewood Cliffs, New Jersey, 1970.

earthquake engineering course without having been able to take that subject as a university student; today's professors teach a course they have already been taught.

Organization of the Berkeley Civil Engineering Department

Bertero: The Civil Engineering Department brought in Bob Park from New Zealand for a short course on reinforced concrete slabs. Park later wrote the book *Reinforced Concrete Slabs* with William Gamble, and of course Park wrote the famous textbook with Tom Paulay, *Reinforced Concrete Structures.* It was very useful to bring in people from the outside to let the Berkeley faculty find out what was going on in other places. We also had the benefit of having short courses and visits from Ferry Borges, who came from Portugal and was their earthquake engineering expert; Emilio Rosenblueth, from *Universidad Nacional Autónoma de México* (UNAM) who wrote the very good book on earthquake engineering with Nathan Newmark, *Fundamentals of Earthquake Engineering*;[31] Jacques Heyman from Cambridge University in England, who was an expert on plastic steel design.

I think the Berkeley civil engineering faculty became so strong partly because the individuals worked so well together as a group and knew how the courses fit together. I don't think I really added significantly to the research capability of the faculty in structural

theory, with all the other leaders in that area like Clough, Penzien, and Scordelis. I think my contribution was more in the experimental area. Some of the apparatus I designed is still used today. In the SAC Steel Project after the 1994 Northridge earthquake, a testing frame I produced years before was used for cyclic testing of beam-column joints, for example. For one project on the shake table in the early years, I spent $30,000 on lead bars to provide the necessary mass to preserve dynamic similitude when the model is smaller than full-scale. Otherwise, the stiffness is too high as compared to the mass and it won't respond realistically.

Reitherman: All those ingots of lead in the Richmond Field Station shake table lab today are from your research project?

Bertero: Yes. Load cells and other instrumentation were also important to have. On the campus in the Davis Hall laboratory, my test setup for testing steel and concrete frames was used for many years. I designed reinforced concrete reaction blocks for testing shear walls in a horizontal position, with prestressing to provide the effect of the gravity load.

I say I contributed mostly in the experimental area, but it was not a question of just testing, but testing with analysis. You must analyze your structure before you test it, as well as afterward when you have the experimental data. This is what I have called Integrated Analytical-Experimental-Analytical Investigations.

I was involved in the full-scale six-story reinforced concrete and seven-story steel buildings at the Building Research Institute in Tsukuba, Japan, which started in 1978. It was called the

---

31. Nathan M. Newmark and Emilio Rosenblueth, *Fundamentals of Earthquake Engineering.* Prentice-Hall, Englewood Cliffs, New Jersey, 1971.

U.S.-Japan Cooperative Earthquake Research Program Utilizing Large-Scale Test Facilities. I directed the dynamic experiments on the largest scale models of these two buildings that could be carried out at the shaking table facility at the University's Richmond Field Station.

International Institute of Seismology and Earthquake Engineering

Bertero: When I had a sabbatical or half-sabbatical in 1971-1972, Professor Hajime Umemura came to Berkeley to teach the course on earthquake engineering. Umemura was the one who followed Kiyoshi Muto at the University of Tokyo.

At that time I went to the International Institute of Seismology and Earthquake Engineering, IISEE, in Japan. Umemura came to the USA and I went to Japan. At the IISEE, they had one foreign professor or expert who taught the courses in the earthquake engineering program and one who taught the courses in the seismology program. Students from developing nations would be paid by the Japanese foreign aid budget to participate in these courses. It was really like a one-year master's program.

Reitherman: In the past few years, IISEE has been accredited to grant official master's degrees.

Bertero: That is a good thing. I found that they were good programs. They were also programs that were designed to serve Japan's interests in having contacts in countries where Japanese companies or agencies can go to develop business and programs, so it works to the advantage of Japan as well. I think the ministry

officials who fund the program regard it as a way to market Japanese products and services to these countries, while the staff who actually run the program see it as a good way to advance earthquake engineering and help the developing nations.

When I was there, the Institute was having some problems obtaining the promise of future funding from the Japanese government. I wrote letters to the minister, but never got any answer. Finally, one of the Institute staff members explained to me that this particular minister did not like Americans from his experiences in World War II, and he was simply going to ignore me. I spent time for nothing trying to help the Institute, until I found out what the story was.

I had a big advantage when I moved there with my wife and my three youngest children, Mary Rita, Adolph, and Richard. My three older children were then in university here in the USA— María Teresa, Edward, and Robert. We were able to move into the residence that Bob Hanson had been renting, buy his furnishings and everything, and just move in. Professor Hanson had been the engineering faculty member there at IISEE just before me. When I was there, I became a good friend of Professor Muto, who gave lectures there.

When I was there, the seismologist teaching that course was an Australian professor, Professor K. E. Bullen, very famous in that field, who had taught Bruce Bolt. Professor Bullen was a seismologist who was not interested in earthquake engineering. He said he was only interested in the properties of the earth. You know, Bob, how much I want to have seismology and engineering combined to solve earthquake problems, so this was not the way I wanted seismology to be taught at the Institute. But he was famous as a seismologist and liked Japan very much and returned several times, and also selected the other seismologists to come to teach. I told him that his former student, Bruce Bolt, was interested in engineering seismology and was doing good work in that field at Berkeley, and he basically said that Professor Bolt was no longer a real seismologist. This has been an issue for many years: most seismologists study earthquakes only as a matter of convenience for the data they provide about the properties of the earth. Only a minority study earthquakes so we can better understand earthquakes.

International Collaboration on Reinforced Concrete

Bertero: One of the accomplishments in my career that I consider to be perhaps the most important one was to be able to direct the organization of the 1977 workshop on earthquake-resistant reinforced concrete, which was held in Berkeley.[32] It was a very large working meeting with the best researchers and practitioners from all over the world. We discussed the state-of-the-art and state-of-the-practice in the seismic design and construction of reinforced concrete buildings, evaluated current progress, and established research need priorities. I remember in the years afterward, when I would run into Bob Park at a conference, he

32. Vitelmo Bertero et al., *Earthquake-Resistant Reinforced Concrete Building Construction: Proceedings of a Workshop Held at the University of California, Berkeley, California, July 11-15, 1977.* University of California at Berkeley, 1978. Three volumes.

would tell me I had to do that again to keep the field up to date.

Reitherman: There are some familiar names here on the title page, besides yourself as the lead. Stephen Mahin was the Organizing Secretary working with you, and the Steering Committee was comprised of William Gates, Neil Hawkins, John Scalzi, Mete Sozen, and Loring Wyllie, Jr. From abroad, you managed to line up quite a blue ribbon list: Shunsuke Otani, Hiroyuki Aoyama, Hajime Umemura, and Toshizaku Takeda from Japan; Bob Park and Tom Paulay from New Zealand; Emilio Rosenblueth and Luis Esteva from Mexico; Ferry Borges from Portugal; Michael Collins from Canada.

Since that important event occurred exactly thirty years ago, [this interview session occurred in 2007], here's a probing question for you: How much change has there been? Did the field progress from a relatively adequate level to an improved level? Or was it a big jump from a low level in the late 1970s to a much higher level of knowledge and practice today?

Bertero: Most of the fundamentals were known. Their implementation in conducting research and in practice had started to improve, but it was not at all to the desired degree. Don't forget that as of 1977, there were really no practical computer programs for predicting three-dimensional inelastic structural behavior under earthquake excitations and therefore for analyzing their inelastic dynamic response. There were still some mysteries then about the shear behavior of reinforced concrete.

Reitherman: What was not known about shear in reinforced concrete as of the 1970s?

Bertero: Diagonal tension was understood, but there was a confusion about the damage that can be produced by bending in combination with changes with axial force, that is, sliding shear. In a shear wall—and I know my colleagues in New Zealand like Tom Paulay would correct me and call it a structural wall— a large overturning moment can put one vertical edge in axial tension, causing a horizontal crack. This is distinct from the diagonal tension causing a diagonal crack. If the horizontal crack is large enough, if the steel reinforcing deforms significantly in the inelastic range, the crack will not close. The steel only recovers its elastic deformation, not its inelastic deformation. The bars have stretched and hold the concrete apart across the horizontal crack. Now consider a load reversal with flexure in the other direction, tipping the wall toward the open crack, and of course, we still have the lateral load to transfer from the wall to the foundation or wall below. What will carry the shear, the sliding shear, across that gap?

Reitherman: It seems very analogous to Paulay's coupled wall analysis, where inelastic tensile deformation in the bars in the flexing link beam upon load reversal keep the crack from closing, and there is nothing but air and the cross-section of the bars to resist the shear. In the coupled wall case, the shear force is vertically oriented and comes from gravity, and the crack is vertical. You turn the coupled wall case on its side and it looks similar to your example of the transfer of horizontal seismic shear across the horizontal crack.

Bertero: Exactly. It has taken time to understand the relationship between axial forces introduced by the lateral response and the

shear that must be transferred. You can handle the problem with diagonal reinforcing, just like Paulay's diagonal link beam method. Even today, I am not confident all our designers understand this. There is a very tall building in San Francisco being built right now, immediately adjacent to the Bay Bridge, with all its lateral resistance contained in its core. As the core walls undergo cyclic loads, one end will experience tension that inelastically stretches the bars, holding the crack apart.

When this sliding shear problem at the base of their core was pointed out, the designers did not want to solve it with diagonal reinforcing. They said it was too expensive. And there will be great congestion of bars at that location, there is no question. But they designed a situation that concentrated the problems there. It gets more complicated when you consider how the core walls will resist torsion, because there is insignificant perimeter frame resistance in that design—again, to save money.

In the 1970s, we understood relatively well the problems of torsion, shear, flexure, axial forces. But we were dealing with them one by one, whereas the actual structure feels them all at once. And from this recent example I mentioned, you can see we have not educated the profession sufficiently.

A virtue of the 1977 workshop was that we brought in practicing engineers so they could explain their design methods and problems and they could also learn the latest research findings.

I was able to contribute a chapter to a book on innovative approaches to seismic design that was a result of an international conference, the Second International Conference on Earthquake Resistant Engineering Structures (ERES), held in Catania, Italy in 1999.[33] After the conference, it was decided to have five or six people contribute to a book that was edited by Professor Giuseppe Oliveto of the University of Catania. I taught several courses in Catania, starting about in 1988. Another of the places where I have taught some short courses in Italy is Napoli, or Naples, and Palermo in Sicily. Most of my teaching of short courses in other countries, however, has been in Latin America.

Loma Prieta Earthquake, 1989

Bertero: I can tell you the story of when I was the Director of the Earthquake Engineering Research Center at the time of the Loma Prieta earthquake, and my secretary came into my office and said, "Professor Bertero, there is a policeman here from the sheriff's office who wants to see you." Ai yi yi! This police officer came to give me the official papers to make me come to a court case concerning what happened when a building in Santa Cruz collapsed in the earthquake.

I was not a designer of any building there, was not a consultant, had not received a penny in any fees, and then I ended up being subpoenaed to appear in the courthouse and was asked all kinds of questions by the lawyers.

I had taught architecture students when I arrived at Berkeley. I didn't remember a former architecture student who came to my office one day, a year before the 1989 earthquake, and said he wanted me to look at a building

remodel he was working on in Santa Cruz. I went to Santa Cruz and took one look at the old brick building and told him he had a serious problem. It was in poor repair. It had poor diaphragms. It had no ties between walls and floors. It was a serious threat to the low wood building next door. I told him falling brickwork could hurt people in the neighboring building. I was not a masonry expert, and I told him he had to get a masonry structural engineering expert and do something. Later he came back with some drawings, and I said, "It's something, but it's not enough." He was trying to use wood elements to provide the retrofit bracing but I said it would not work. I asked him why he wasn't doing more and he had various explanations. He said he couldn't do anything about it right away. There were various explanations, but nothing was done.

Then, when the Loma Prieta earthquake occurred, October 17, 1989, I sent two of my doctoral students at that time, Andrew Whittaker and Eduardo Miranda, out to look at the damage. Andrew went to report on the Cypress Viaduct. Eduardo went to Santa Cruz. Eduardo came back and reported the sad fact that the brick building had partially collapsed, and bricks had fallen on the lower building next door, the Coffee Roasting Company, and killed three people. What a tragedy! And it could have been prevented.

Sometime after the earthquake was when the sheriff's deputy arrived with the subpoena for me, telling me I had to be a witness in a lawsuit. It was a very complicated lawsuit, a triple suit. The owner of the building that the bricks fell on sued the owner of the brick building. The relatives of the people who were killed sued

---

33. Vitelmo Bertero, "Innovative approaches to earthquake engineering," *Innovative Approaches to Earthquake Engineering*. Giuseppe Oliveto, ed., WIT Press, Southhampton, UK, 2002.

Table 2: *Visiting Scholars at Berkeley With Whom Bertero Collaborated*

Aktan, A. E.	Lara-Montiel, Otton (Ecuador)	On the recommendations of Professors Bertero, Bresler, Clough, Penzien, and Popov, these visiting scholars were supported by the U.C. Berkeley Division of Structural Engineering and Structural Mechanics to give a series of lectures or short courses: Julio Ferry Borges (Portugal), Emilio Rosenblueth (Mexico), Hajime Umemura (Japan), and Robert Park (New Zealand).
Aktan, H. M.	Liao, W.-G. (Taiwan)	
Anderson, J. C.	Linde, P. (Switzerland)	
Bertero, R. D. (Argentina)	Llopiz, C. (Argentina)	
Bonelli, P. (Chile)	Lobo-Quintero, W. (Venezuela)	
Celebi, M.	Mollaioli, F. (Italy)	
Del Valle Calderon, E. (Mexico)	Oliveto, G. (Italy)	
Eligenhausen, R. (Germany)	Ozaki, M. (Japan)	
Endo, T. (Japan)	Rodriguez, M. (Mexico)	
Filiatrault, A. (Canada)	Sakino, K. (Japan)	
Giachetti, R. (Italy)	Santana, G. (Costa Rica)	
Gonzalez, G. (Colombia)	Sugano, S. (Japan)	
Guevara-Lopez, T. L. (Venezuela)	Vulcano, A. (Italy)	
Harris, H. G.	Watabe, M. (Japan)	
Igarashi, I. (Japan)	Yoshimura, J. (Japan)	

the people who owned both buildings. And I had to go to court for several days. I told the court that I had told the owner of the building before the earthquake that engineers should be retained to do a thorough retrofit, but it wasn't done. They kept asking me all kinds of questions, for several days. I had never received one penny, but I spent a lot of time in court.

Teaching Architects

Bertero: The very first course I taught at Berkeley in the fall of 1958 was statics for the architecture students, and after that I taught them strength of materials. My first course was a big, big surprise. It was hot weather when we started the semester in August, and the female students were wearing shorts that were very short, and sandals rather than shoes. The male students were dressed very casually. MIT was not like that. And in Argentina, the male students all wore coat and tie, and the female students would never have worn what they were wearing at Berkeley. It was a shock for me. I said to myself, well, this is America, and more particularly this is Berkeley, and I have to become acquainted with these customs.

At Berkeley, I don't think the architecture students take as much engineering as they used to, and the engineering students don't take architecture and urban planning courses. I don't think we have enough collaboration between architects and engineers today. And the engineers need to understand why the architects want to design the buildings they do, and this should start in their undergraduate studies.

I have worked with a Venezuelan architect, Teresa Guevara-Perez, whose husband is also an architect. In a book they are writing, her husband points out why some seismically vulnerable configurations are used for buildings. It is not that architects are trying to make earthquake problems, they are just dealing with other concerns and not paying enough attention to earthquakes.

Take for example the setback. It introduces structural discontinuities at each setback level. But there are architectural and urban planning reasons for them. What do you call the famous New York law about zonation?

Reitherman: That would be the New York zoning law of 1916. It became a model for requiring setbacks, so that as tall buildings went up, they had to step back, or set back, from the perimeter at the base—getting smaller toward the top. Just before that, there was a tall building, the Equitable Building in Manhattan, that went straight up forty stories from the sidewalk, like one big rectangular prism. Such a shape blocks a lot of light and views. There was no thought of earthquakes in New York when they passed the zoning law; it was for urban planning reasons.

Bertero: In Latin America, often there is a legal way to build more floor area at the upper levels with projections that don't count as the outline of the building. So there are large balcony structures. It doesn't look rational from a structural viewpoint, but because they are balconies, they fit within the zoning rules.

Another example is the architect Le Corbusier and the use of pilotis—the idea of raising the building off the ground on ground-story stilts.

He also had designs for tall buildings where there was a story at mid-height where the walls above and below stop and there are just columns, which I think was so people could walk around there like a plaza except high off the ground and enjoy the view. But when you do that, you introduce a structural discontinuity, the stories with the walls being much stiffer than the story with the columns. This caused some problem configurations in buildings in Algeria, which is highly seismic, when it was a French colony. You must have communication between the architects and engineers.

Reitherman: How did you like teaching the architecture students?

Bertero: I liked it. Then they started to remove the requirement for the architecture students to take structural engineering classes. Karl Steinbrugge, a structural engineer, was a professor in the architecture department, teaching structures to the architects, and he supported me in trying to keep more structural engineering education for the architects.

Reitherman: An architecture professor at Berkeley recently noted that today structural engineering classes are recommended but not required for the architecture degree, beyond one survey course on structures.

Bertero: In Argentina, I also taught the architecture students the design of structures. In 1953, when I told them that I was leaving to go to MIT to study and do research, they expressed their thanks for my teaching and signed a nice diploma or certificate, where they expressed their best wishes for my trip and studies in the U.S. It was a very nice thing for them to do.

It is sad that the architects start the whole design process, but are not really interested in seismic design. If you look in the EERI directory, how many members are in the architecture category?[34]

Reitherman: When I finished my master's in architecture at Berkeley in the late 1970s, I was fired up about seismic design from studying with Karl Steinbrugge and others there. I quickly found that there was an engineering market for that kind of expertise, but not in the architecture profession. Even with regard to nonstructural components, such as ceilings, partitions, and glazing that are within the architect's scope, you'll find almost all the research and consulting work in the earthquake field is done by engineers, not architects. That's why "earthquake engineering" is a well-known term and "earthquake architecture" has a strange sound to it.

How would you try to solve the problem of lack of involvement of architects in solving earthquake problems?

Bertero: The universities should teach the architecture students about earthquakes. It would be less oriented toward the precise rules of the building code and the calculations, but it would explain how the architectural design decisions affect the seismic design.

It would also be good if the architecture and engineering students could work on some joint project at the end of the semester.

---

34. Approximately thirty members of EERI are listed under the architecture disciplinary category at present [2008], out of approximately 2,500 total individual members, or about 1 percent of the membership.

A way to educate the practicing architects would be to arrange for them to visit the scenes of earthquake damage. Most earthquake investigators and students visiting the scenes of earthquakes are engineers, earth scientists, or social scientists. Giving architects a first-hand view of what happens in an earthquake is a great opportunity.

Reitherman: At Berkeley, the heyday of involvement of architecture faculty and students in earthquakes was probably the 1970s. Four civil engineering structures courses past the first introductory architecture course on structures were required. And not only was the structural engineer on the architecture faculty, Karl Steinbrugge, an earthquake expert, but architects on the faculty like George Simonds, Henry Lagorio, Gerald McKue, and Chris Arnold were also active in the field. Eric Elsesser was a lecturer in structures in the architecture department then also. Nationally, the research arm of the American Institute of Architects (AIA) in Washington, AIA Research Corporation, had several seismic projects in that same era, and the president of the AIA in 1978, Elmer Botsai, wrote some things on architects and earthquakes and took an interest in the subject.

Bertero: I have enjoyed collaborating in the earthquake field with architects, such as Chris Arnold, Mary Comerio, Henry Lagorio, and Teresa Guevara-Perez.

Early Teaching at Berkeley

Bertero: I only taught one course the first semester I was at Berkeley, the statics course for architecture students, but I was also given the assignment of helping a consulting engineer teach the course on earthquake-resistant design of wood structures. I knew his name because of his participation in the 1956 World Conference on Earthquake Engineering. It was Henry Degenkolb. Henry had designed many wood buildings for the 1939-1940 Golden Gate International Exposition on Treasure Island in San Francisco Bay, so the university had him teach that subject. I became good friends with Henry. He gave the students lots of practical problems on the design of wood structures, and I would meet with the students and go over their questions. Henry did a lot to improve the engineering profession. I knew him from then on, till he passed away in 1989, and collaborated with him in inspecting damage after significant earthquakes.

I was very lucky from the start at Berkeley, with Joe Penzien and Henry Degenkolb, and of course with my discussions with Professor Popov regarding research activities.

I was also supposed to assist Professor Howard Eberhart in the structures laboratory, preparing the laboratory setup for the students and help in the lab sessions. Professor Eberhart assigned the students who were not doing very well to himself, and the better students were assigned to me. Oh my! He worked the students in his section hard.

Then I helped Professor George Troxell, who was testing the theory of an engineer in the Office of Architecture in Sacramento, the people who administered the Field Act, that drilling holes to remove the knots in lumber increased its strength. It was an unusual project. It turned out that sometimes it might help to take out the defect that would initiate a split, but sometimes it didn't help, or it reduced the section too much and made it worse, such as when you put a notch near an edge, which only made a stress concentration.

Tenure

Bertero: After I was at Berkeley for two years, having started in 1958, I received a letter from Bob Whitman asking if I was interested in returning to MIT for a position in the soil mechanics area. That would have made two structural engineers becoming geotechnical engineers there at MIT, but it did not happen. I did not yet have tenure at Berkeley. I talked to my fellow faculty members and they advised me not to rush into the MIT opportunity. They said they were already recommending me for tenure. Normally, the decisions to promote someone to associate professor were made over the summertime. But I think the university hurried up a little bit to give me my tenure early to keep me from leaving, so it was in May of 1960 that I received this letter [which Bertero pulled from a file; it is a typed letter signed by Clark Kerr, then President of the U.C. system]. It is dated May 31, 1960, so in other words that was a little bit early to get this kind of appointment. I decided to stay at Berkeley.

Research on Steel Connections With Egor Popov

Bertero: After some of the little research things I was assigned when I began at Berkeley, the first major research project I worked on was with Professor Egor Popov. The topic was the cyclic behavior of steel joints. We got very good results. They were as I had expected. If you let the flange of the column or beam go inelastic too much, then as you cycle the load you have a buckling bulge or curve when the flange is loaded in compression; and it curves more and the strain along that curving region becomes tremendous. Remember that as the material reaches that flat portion of the force-deformation curve, it has very little buckling resistance. It's like taking an arch and squeezing it at the abutments, making the curvature increase. You can get a complete fracture in as little as five or ten cycles at 2 percent strain. If you didn't have that inelastic behavior, you might have 500 or 1000 cycles before you had a fatigue failure. This was an earthquake engineering research project, because that is the only way you would get several cycles with large deformations.

I designed and built the instruments to measure the strains at different places on the steel specimens that we were testing. It was based on similar devices that had been used at the Torroja Institute in Spain. It was like a tiny three-legged table, with the base of each leg made of a steel phonograph needle that sat in a tiny hole in the surface of the steel. Those formed pin joints. There were rigid connections between the legs, which were very stiff, and the "table-top," which was a thin phosphor bronze beam that had strain gauges mounted on it. If you changed the distance between the legs, you flexed the tabletop, accentuating the strain for better measurement. It worked quite well.

Professor Popov and I wrote a short paper on this research that took four years to publish.[35]

35. Vitelmo V. Bertero and Egor P. Popov, "Effect of Large Alternating Strains of Steel Beams," *Journal of the Structural Division, Proceedings of the American Society of Civil Engineers*. Vol. 91, February, 1965, ST1, p. 1-12.

I can explain to you why. Other important steel researchers, such as at Lehigh, as well as some practicing engineers, thought that our results could only happen in the laboratory, that it was not a design issue, and that we were going to scare the steel industry. Then in 1964, the Alaska earthquake occurred. I went there with Professor Joe Penzien. Oh, it was so cold! We walked around and learned a lot. Soon after that, I think Joe became more interested in analytical topics than looking at earthquakes, but then we were out looking at earthquakes. We looked at the Cordova Building in Anchorage, and one of the steel columns looked as if it had been cut with a torch. Other steel columns in the Cordova Building were also damaged, with buckled flanges. It was the kind of failure Professor Popov and I had obtained in our experiments in the laboratory. After the earthquake, we were able to get our paper approved for publication. That's why it took four years.

The old K Factor in the Uniform Building Code, the factor for the structural system, really was a way of considering the structure's ductility. If you don't detail the structure for the expected range of ductile behavior, then it should be reflected in a different K Factor—that is, a higher factor in the base shear formula that will result in a higher level of design forces. So it was a significant piece of research, and the steel industry and codes had to adjust the seismic design procedure for steel structures to maintain a favorable K Factor. Without changes in steel design, the K Factor would have increased and that would have increased the cost of steel buildings.

Director of the Earthquake Engineering Research Center

Bertero: In 1988, I was asked to take a term as Director of the Earthquake Engineering Research Center (EERC) at Berkeley. I ended up being director from then until I had my back surgery in 1991.

They politely told me, "Vitelmo, you have to do this. EERC needs a director and you should do it." Oh my! It practically spoiled my career. I said I was not an administrator, that I was a researcher, a teacher, an engineer. But I did it.

At that time, EERC had been in operation since Joe Penzien had been the first director in 1968. When I was director, twenty years later, there were some staffing problems. We were not getting enough projects through the lab, and it was taking too long to finish the upgrading of the shake table and put it back into operation. Briefly, after I retired, the university brought in Bob Hanson to be director. After Bob was Jack Moehle, and currently [2007] the director is Nicholas Sitar.

I would go home at night and say to my wife that I was unhappy, saying that I had solved one problem but didn't solve all the problems, and there would be new problems the next day. Administration? Bureaucracy? I cannot do a good job with those things.

Reitherman: Don Clyde, the senior research engineer at EERC for many years, has said that you instituted Friday morning meetings of the staff to get more control over who was doing what. He said you really made people perform, even if they hadn't been used to it. [At this point in this interview session, Chuck James, EERC Librarian, entered Bertero's office to ask him to say a few words at the retirement party later that day, May 2, 2007, for Ruth Wrentmore.]

Bertero: Speaking of employees, when I was EERC director, Ruth was the best. She was always at work on time, always reliable. Ruth was the one who started the Earthquake Abstracts series and maintained it over the years. Her name then was Ruth Denton, now Ruth Wrentmore. She compiled the abstracts from the literature in the days when it was done on the typewriter, using paper of different colors to categorize them. Then, she was the one who oversaw the transition of the program to a computerized system. In her tenure running that important program, there have been about 140,000 pieces of literature cataloged.

Political Protests of the 1960s and 1970s

Reitherman: What became known as the Free Speech Movement began on the Berkeley campus in the 1964-1965 school year. Did the political activity of the 1960s permanently change the university, or did it cause only temporary changes?

Bertero: Before the protests of the 1960s, the university students had less voice in campus policies. Now they do. So that is a big, permanent change. Perhaps it was a mistake for the university not to give the students more influence earlier.

The Civil Engineering Department was not very much affected. If you put a line down the middle of the campus through the Campanile, from the hills in the east to the west toward the

Bay, you had two different universities. The humanities, the design college with the architecture department, and other departments on the south side of the line were more involved in the politics of the time. The engineering college, computer science department, and sciences, which were on the north side, were not as affected.

I remember one day having lunch in the Faculty Club, and there was a student demonstration. The police and national guard came, there was tear gas, and we couldn't leave the building.

I remember also the People's Park controversy. I never really understood that issue. I think the result was that instead of student housing being built, the block ended up somewhat like a vacant lot.

Changes in the Civil Engineering Department

Reitherman: In your long tenure at the university, what has changed the most?

Bertero: In civil engineering, when I started at the university, we had divisions within the department devoted to transportation, structures, hydraulics, and surveying. Then a construction division was added, but the department faculty still worked together.

But when it became the Department of Civil and Environmental Engineering, it was detrimental to communication within the department. Before, there was a goal of educating the civil engineering student in a broad background in the field. The person who specialized in structures also had to know about materials, for example. But now, the subject matter

is so large it is hard to cover everything. The environment is a vast subject. It is not possible in four years to give the student the same basic education in the previous civil engineering subjects that used to be taught. The student must specialize in a smaller fraction of the whole field, or get a more general education. And the faculty in such a large department that has so many disciplines does not know one another and work together as we did. As I mentioned earlier, the rise of U.C. Berkeley in the earthquake engineering field occurred not just because there were some good individual faculty members. It happened because we worked together.

Reitherman: In the publications of the practicing structural engineers, the call is for more specialization of students in structural engineering. You don't see many practicing engineers calling for a broad university education of engineers, but instead they are calling for what might be called vocational preparation. That is perhaps no surprise, since practicing engineers are the employers and young graduates are the employees. I spoke with the president of one national structural engineering organization who kept referring to the "product" the universities were producing, and it took me awhile to realize that "the university product" was, to him, synonymous with "the university graduate."

Bertero: You can produce a good technician that way, but you cannot produce a good designer. You can produce graduates who will be productive when they are first hired, but their university education is all they have to build on as a foundation as they add knowledge on the job. Their university education must give them a broad background. Earthquake problems cannot be solved by engineering alone. The engineers need to know the other aspects of the problem.

Retiring From Teaching

Bertero: I loved my teaching, but I developed a serious back problem that made it very painful for me to stand up, to raise my arm to write on the blackboard. I needed to have back surgery, and it took several months to recover from it. That was in 1991, and I retired from teaching then. I still do the research and some short courses and lectures, but not the teaching of regular university courses.

Classes started at ten minutes past the hour, such as 10:10 am. I would go in the classroom at 10:00 and write my diagrams and equations on the blackboard. The students would come in later, and sometimes they thought they were already behind when the class started. But I thought it was more efficient for me to take the time in advance to put the material on the blackboard, so I could teach more in the following fifty minutes.

Chapter 8

Working With Practicing Engineers

I have tried very hard to provide the practicing engineers with what they need to do a good professional job.

Bertero: When you said you were going to ask me in this interview session about practicing engineers I have worked with, I prepared a brief list (Table 3).

Reitherman: You're known as one of the university professors who has had the closest connection with professional engineers, and who has been most concerned about the application of seismic design knowledge in practice. When there was a symposium in your honor, for example, the large turnout included a high percentage of practicing engineers.[36]

Allow me to read into this oral history some of the other honors you have received from practicing engineers. In 1990, *Engineering News-Record* made you their Man of the Year. The select few people who earn that honor have made big contributions to the practice of engineer-

---

36. *EERC-CUREe Symposium in Honor of Vitelmo V. Bertero,* January 31 February 1, 1997, Berkeley, California. Earthquake Engineering Research Center, 1997.

ing. It is not an academic type of award, it is only given to people who have had an effect on the construction and engineering industry. I think Henry Degenkolb and John Blume are the only other ones to make the cover of *ENR* whose career was focused on earthquake engineering. In 1997, the Structural Engineers Association of California made you a Fellow. In 2006, during the week of the commemoration of the centennial of the 1906 San Francisco earthquake, you were given an award by *Engineering News-Record* and the Applied Technology Council as one of the top thirteen American earthquake engineers of the twentieth century.

Bertero: I am grateful the practitioners have appreciated my work. I have tried very hard to provide the practicing engineers with what they need to do a good professional job. I have had that goal from the first time I started teaching and doing research. When I became director of the Earthquake Engineering Research Center at U.C. Berkeley and had so many administrative tasks, all my activities with the practicing engineers virtually disappeared, because I had no time, and that was not good.

It is important for the professor to have some practicing engineering background. What does a professor teach? Most of your students are going to practice engineering, they are not going to be researchers. How can you teach the practice of engineering if you haven't practiced engineering? I don't think it is a good trend when young people go directly from high school to college to PhD to being a university professor.

I would say the first time I tried to make a contribution to the practicing engineers was in

1964, when I worked with a professor by the name of Herbert A. Sawyer, Jr. on the International Symposium on Flexural Mechanics of Reinforced Concrete. That was really a way to present the theory of limit state design to the practicing engineers. It required years of education and familiarization with that idea for it to be adopted in codes and in practice. It was one of the big changes in structural engineering in the last half of the twentieth century, and it required a big effort to educate the profession, as well as educate the future engineers who were in college then.

The steel people lagged behind the concrete people in that process of adopting new ways. The concrete people were quicker to adopt concepts of inelastic seismic design, which the steel field did not fully accept until 1997.

ATC 3-06

Bertero: After that, I would say another significant contribution was when I worked on the ATC 3-06 project.[37] A very large number of practicing engineers, as well as a smaller number of professors, were involved. Many prominent consulting engineers were involved. It was a tremendous experience.

I noticed then the difference between the practitioner and the academic. The practitioner wanted simplicity, and I understand that. I will never forget hearing the heated discussion about how complex the inclusion of dynamic

---

37. Applied Technology Council, *Tentative Provisions for the Development of Seismic Regulations for Buildings.* ATC 3-06, 1978. Funded by the National Science Foundation and the National Bureau of Standards.

theory should be. Henry Degenkolb said that to dance the tango you need two things: you need the theory, but you also need the practice in order to know what a practicing engineer needs. It was not always a debate between academia and practice. Egor Popov and Boris Bresler were examples of professors who practiced and bridged the two. Nathan Newmark, the key leader of the project, did a tremendous job cooling things down when necessary, balancing the committees, and keeping the whole group together. Because of Newmark, two younger professors at Illinois, Bill Hall and Mete Sozen, were involved, and Newmark also was responsible for Emilio Rosenblueth coming from Mexico for the meetings. Sometimes Hall and Sozen couldn't make the trip to San Francisco for a meeting, and I would find myself outvoted in our committee where the R Factor was being developed. As we can discuss later, to this day I have major reservations about the R Factor [see Chapter 10, "Economic Pressure to Select Structural Systems"]. Roland Sharpe was then the executive director of ATC and did an excellent job also. For a young professor such as me, it was not just a question of making a contribution; it was also a tremendous benefit to learn so much from others.

The only thing that was always a problem in my mind was how long it would take to finish the project and see the results implemented. We finished our work in 1975. The thick report was published in 1978. But it did not immediately affect the code. It takes time to build a consensus in support of change. You cannot change the profession overnight.

At that time, ATC was a small organization that was set up by SEAOC. Its main purpose was really to translate the earthquake engineering work of academics so practitioners could stay up to date. Later, ATC tried to get big projects and do more than be a bridge to the profession, which brought some positives and negatives compared to the era of ATC 3-06.

Reitherman: Joe Nicoletti was one of the engineers who had the task of going out from California, where ATC 3-06 was basically developed, to try to get national support for it. In his EERI oral history, he recalls being at a meeting in Maryland organized by the National Bureau of Standards to bring together all the materials and trades organizations, the model code organizations, to hear about ATC 3. He says that out of about sixty people in the room, other than the dissent of the ATC presenters—himself, Ron Mayes, and Rol Sharpe—there was a strong consensus. And that consensus was completely against the new proposed provisions.

Bertero: I recall lecturing on the East Coast after ATC 3-06 was published and getting the same reaction.

Reitherman: ATC 3-06, thirty years later now, still has some advanced material in it. For example, in Chapter 1 of the Commentary, it has a discussion of the overall probabilistic intent of the provisions. By "overall" I mean it tried to combine all the uncertainties to bluntly answer the question: If you use these provisions to design 100 buildings, and if the design earthquake that has a particular chance of occurrence materializes, how many of those buildings will fall down? What if the earthquake motions are twice as intense as the design-level motions? And is that risk the same across the different seismic hazard levels or map areas? It was an unusually clear presenta-

Table 3: *Selected Practicing Engineers in the USA With Whom Bertero Worked*

H. Bauman	Ron Hamburger	Chris Rojahn
K. L. Benuska	Bill Holmes	Charles Scawthorn
John Blume	Ron Mayes	Roger Scholl
Boris Bresler	Frank McClure	Dan Shapiro
P. Crosby	McCreary-Koretsky Engineers	Roland Sharpe
Henry Degenkolb	Jack Meehan	J. P. Singh
Eric Elsesser	Farzad Naeim	Charles C. Thiel
Eduardo Fierro	Joe Nicoletti	Tom Tobin
Nick Forell	Cynthia Perry	Loring Wyllie, Jr.
Sig Freeman	Clarkson Pinkham	Peter Yanev
William Gates	Chris D. Poland	Nabih Youssef
Ben Gerwick	Robert Preece	Ed Zacher

tion of the overall risk. I have heard that Jack Benjamin had something to do with that.

Bertero: And it also came from Rosenblueth and Newmark. Their book, *Fundamentals of Earthquake Engineering*,[38] was published a few years earlier, and is still today a good textbook. Rosenblueth added a strong element of probabilistic thinking to that textbook.

Working With Practicing Engineers

Reitherman: Let me ask you about some of the names on your list (Table 3). Do you recall the first practicing engineer with whom you collaborated after arriving at Berkeley in 1959?

Bertero: I mentioned earlier that one of the lecturers I was assigned to assist when I arrived at Berkeley happened to be Henry

Degenkolb, who was teaching a course on the earthquake-resistant design of wood structures. So he was the first. That was in the academic setting, however. I learned a lot from Henry regarding the practice of earthquake engineering. Later, in code-related meetings, I saw how Henry was in some ways conservative, which can be a good thing when you really worry about changes in design approaches that could result in construction getting built that might be worse, not better. Degenkolb was very blunt in saying what he thought, which took a little getting used to.

Reitherman: Frank McClure's name is on the list. How did you meet him?

Bertero: He took a short course on dynamics from me one summer in the 1960s. After he practiced with David Messinger, he left to work for the University of California at Lawrence Berkeley Laboratory. After the 1985 earthquake in Mexico, I worked closely with him on EERI activities when he was president.

38. Nathan M. Newmark and Emilio Rosenblueth, *Fundamentals of Earthquake Engineering*. Prentice-Hall, Englewood Cliffs, New Jersey, 1971.

Reitherman: What about these two related names, Nicholas Forell and Eric Elsesser,[39] the partners of Forell-Elsesser Engineers?

Bertero: Eric has always been looking forward, thinking about how to improve the structural engineering and architectural design professions, particularly as related to seismic design. I met Eric at U.C. Berkeley. Did you know he was a lecturer there in the architecture department?

Reitherman: I had him as an instructor there, in the master's degree working drawings class. Very creative and inspiring. He sketches and thinks like an architect, but also has the quantitative, analytical element of intelligence to quickly resolve the essence of an engineering problem.

Bertero: Eric and I started working together on some of the problems related to the improvement of the education of architectural students in the area of structural design, already in the late 1960s and then in the 1970s, during the ATC 3-06 project. As for Nick Forell, I first knew him after the El-Asnam earthquake in Algeria in 1980. Haresh Shah co-edited that EERI reconnaissance report with me.[40] I was good friends with Nick after that. Nick was also a tough-minded engineer, like Degenkob, but when he explained why he disagreed, he had a very gentlemanly manner. After the 1985 Mexico earthquake, a small group of us got together to discuss the earthquake and early performance-based design ideas, including Forell, Elsesser, Chris Arnold, Sig Freeman, Nabih Youssef, and Rol Sharpe. The Vision 2000 effort of SEAOC grew out of that little discussion group. Forell-Elsesser then hired a young PhD of mine, Andrew Whittaker, in the late 1980s. Whenever I would see Nick Forell, he would say, "Vitelmo! You sent me the most energetic engineer in the world! I'm too old to keep up!"

Forell-Elsesser Engineers has always been willing to explore the most innovative ideas. The firm produces reliable designs, but they explore what is new.

Reitherman: John Blume is another prominent earthquake engineer on your list.

Bertero: Blume was an innovator. He was highly educated and in his practice dealt with unusual problems. He was one of the five founders of EERI. Blume had already published in 1936 a paper on his forced vibration testing to learn about the dynamics of structures for purposes of seismic design.[41] I became familiar with his work at the end of the 1950s. In 1956 at the World Conference on Earthquake Engineering, he presented an important paper on his work on a fifteen-story building in San Francisco that he had studied over the years, going back to his master's degree days at Stanford.[42] Another of his early papers was

---

39. Eric Elsesser passed away in 2007.

40. Haresh Shah and Vitelmo Bertero, *Preliminary Reconnaissance Report: El-Asnam, Algeria Earthquake, 10ᵗʰ October, 1980.* Earthquake Engineering Research Institute, Oakland, California, 1983.

41. John A. Blume, "The Building and Ground Vibrator," *Earthquake Investigations in California, 1934-1935.* Special Publication No. 201, U.S. Coast and Geodetic Survey.

on his reserve energy concept.[43] That's another example of how John Blume was ahead of his time. In the mid 1970s I worked with him in the Seismic Design Review Group of the ATC 3 project team. Among the other members of that committee on design were Professors Housner, Newmark, Whitman, and Clough. As I mentioned earlier, Dr. Blume helped me in the organization of the Workshop on Earthquake-Resistant Reinforced Concrete Building Construction, and he delivered the keynote lecture there on an overview of the state-of-the-art.

As you know, John Blume headed up the consulting firm of John A. Blume Associates, later called URS/Blume, and he and the engineers there published many important papers and implemented advanced seismic design in their projects. Earlier I explained that first there was "earthquake-resistant construction," the emphasis on rules of thumb with little engineering. As civil engineering developed in the second half of the twentieth century, the field evolved to "earthquake engineering," the application of modern principles of mathematics, applied physics, structural engineering. And in that evolution, John Blume was a very important person.

---

42. John A. Blume, "Period Determinations and Other Earthquake Studies of a Fifteen-Story Building," *Proceedings of the World Conference on Earthquake Engineering*, Berkeley, California, June 1956. Earthquake Engineering Research Institute, Oakland, California.

43. John Blume, "A Reserve Energy Technique for the Design and Rating of Structures in the Inelastic Range," *Proceedings of the Second World Conference on Earthquake Engineering*, Tokyo and Kyoto, Japan, 1960. International Association for Earthquake Engineering, Tokyo, Japan.

Reitherman: Another engineer in the EERI Oral History series is Clarkson Pinkham, who is also on your list. Did you get to know him in the ATC 3-06 project?

Bertero: Yes, but even more in the U.S.-Japan Cooperative Earthquake Research Program Utilizing Large-Scale Test Facilities, during the phase devoted to steel. He was always a good contributor.

Ed Zacher is another name on the list of practicing engineers I recall, who perhaps was never recognized enough for his contributions. He had a very good feeling for the physical phenomena. Quite a gentleman. Very nice to work with him. SEAONC has an award now in honor of him, the Edwin G. Zacher Award.

Bill Holmes has worked on many earthquake engineering projects over the years, but I got a chance to work closely with him on the U.C. Berkeley Disaster-Resistant University Project, which Mary Comerio of the architecture faculty at Berkeley managed. There were several consulting firms involved—Rutherford & Chekene, Forell-Elsesser, Degenkolb. The goal was to realistically predict the actual behavior of the buildings. Bill had a good understanding of ductility. One of the best consulting engineers.

Chris Poland is an interesting fellow. You know, he has an undergraduate degree in mathematics, not engineering. I got to know him well in the EERI reconnaissance effort for the magnitude 7.4 1977 Caucete earthquake in Argentina. I worked with him when he was the chair of the SEAOC Vision 2000 project also. He is a very good engineer and sees the problems of the future. He's pushing a concept called resilience. I use the term to describe a mechanical characteristic, and I think he uses the term differently—to describe how to recover quickly from an earthquake. As president of the Degenkolb firm, Chris has become an executive who is good at getting the job done.

Reitherman: Chris is the EERI president who authorized the current phase of the EERI oral history program after Stan Scott passed away in 2002. He said if I had the will to continue the program, he would support that. I checked in with him occasionally and he would just ask if he needed to clear anything out of the way for me and the new oral history committee. He steers a straight course.

Bertero: Exactly. Chris knows what has to be done in order to get it done.

Another outstanding engineer with Degenkolb Engineers is Loring Wyllie. Of course, there was Henry Degenkolb himself, whom we talked about earlier. I met Loring when I was teaching at Berkeley and he was an undergraduate student. Like Chris, he is another former EERI President. He can be like Henry Degenkolb—very strong in sticking to the course he sets. Henry and Loring have achieved a great deal by having that determination over the years. Loring was an excellent chair of a peer review panel on tall buildings of which I was a member.

Reitherman: You have Ron Hamburger and Peter Yanev on your list. Tell me about them.

Bertero: When Ron came to San Francisco to work for the firm called EQE that Peter Yanev had founded, I don't think he had much background in earthquake engineering. But once he was here, he learned so much. I knew

Peter Yanev back even further, when Peter was a student getting his civil engineering degree at Berkeley. After Peter obtained his undergraduate degree from Berkeley he got his MS at MIT, then worked for the Blume firm. He was a nephew of Professor Frank Baron, a colleague and good friend in the Structural Engineering and Structural Mechanics Division at U.C. Berkeley. Peter hired Ron here in San Francisco after Peter established EQE. I was on a peer review panel with Helmut Krawinkler, Joe Nicoletti, and Egor Popov looking at a building EQE was going to retrofit. Ron was very receptive to our critique. He has been able to learn very quickly. When Frank Heger from Simpson, Gumpertz, and Heger called me, whom I knew from our old MIT days, I recommended that they recruit Ron to join their San Francisco office. At that time, EQE had been bought by ABS, and I think it became less satisfying for the structural engineers doing design at EQE. Ron knows modern seismic engineering and can also explain it to others. He's a bit like Chris Poland in knowing how to manage projects and also be a designer.

Reitherman: T. Y. Lin was of course a fellow faculty member with you, but you also list him as a practicing engineer with whom you collaborated. Would you say a few words about him as a designer?

Bertero: I was a consultant to T. Y. on some of his design firm's projects. One is the reinforced concrete frame building in Emeryville by the Bay Bridge, Y-shaped in plan, just inland from the freeway. I think it was the tallest reinforced concrete frame building in the highest seismic zone in the U.S. Ray Clough ran the analysis, and I was asked to consult on the detailing. It was a difficult problem to detail the extensive confinement needed. I remember when we hired Jack Moehle to the faculty he looked at the building when it was under construction and he asked if it was a concrete building or a steel building! It was important to have all that reinforcement. T. Y. was able to make design decisions very quickly, after listening to the design team. He had that intuitive sense of design in his prestressed concrete structures. He needed that creativity to come up with the Ruck-A-Chucky Bridge design. The Ruck-A-Chucky Bridge in California was never built, but it was a very innovative design to use cables to support a bridge deck that curved in plan as it crossed a canyon.

Reitherman: As a professor giving a lecture on seismic design at Berkeley, T. Y. Lin would not only narrate the difference between the response of a higher-frequency, short building to an earthquake as compared to a tall, low-frequency one. He would shake his slight frame vigorously to illustrate the former and then undulate his body gracefully and make his fingers move fluidly to demonstrate higher modes for the latter. I learned later he and his wife were accomplished dancers and had a ballroom dance floor built into their house, which matched up with my memories of how he carried himself in front of the classroom.

Bertero: He gave an exhibition of ballroom dancing for a big party at his house, and he was really something!

Reitherman: Talk about another engineer on your list, Bill Gates.

Bertero: Bill was an early graduate student of mine. At the time of the San Fernando

earthquake, he was one of the few engineers who could do dynamic analysis on the computer to compare the analytical response with the measured response from the strong motion records. Later he worked for Dames & Moore in Los Angeles. The firm later became part of the URS Corporation. Bill has been a very methodical and successful structural engineer. In 1977 he was a member of the Steering Committee of the Workshop that I organized on Earthquake-Resistant Reinforced Concrete Building Construction.

Reitherman: Did you meet Charlie Scawthorn when he was also working for Dames & Moore, in their San Francisco office?

Bertero: Yes. He happened to live only a few blocks from my house in Berkeley. We worked on an EERI committee for several years in collaboration with Japan on the seismic retrofit subject. That was when the Japanese thought the next big earthquake would be in Shizuoka Prefecture. You know, Charlie got his doctoral degree from Kyoto University. I think he has always had a sentimental feeling toward helping the Japanese. He has been very successful in doing so as a practicing engineer and later as a professor at Kyoto University. He has a broad knowledge of modern seismic engineering and is a good writer.

Reitherman: You've mentioned ATC a lot, and Chris Rojahn is on your list. How did you meet him? Chris has been the executive director of ATC since 1981. Did you know Chris before his days at ATC? He was a strong motion engineer at USGS when the 1972 Nicaragua earthquake occurred, well before his time at ATC.

Bertero: Yes, I had some contact with him when he was at USGS prior to his arrival at ATC. I worked with Chris as a member of the EERI reconnaissance team for the 1977 Caucete earthquake in Argentina. Chris followed Rol Sharpe and Ron Mayes as Executive Director of the Applied Technology Council. From his work in strong motion research, he went on to become a very good administrator of the ATC organization. The Applied Technology Council has grown tremendously under Chris's direction. The project name ATC 3 indicates it was the third project for ATC, and the number of its projects is now up over seventy, and many are multi-year, large projects involving many people and several reports. Usually, an ATC project has one or more technical leaders of a project and a review committee, and I have served on several of those committees.

Another disaster outside the earthquake field—the attacks of September 11, 2001—diverted attention of FEMA from earthquakes and has slowed down one ATC project, ATC 58, that I would like to be completed as soon as possible. Its goal is to produce performance-based design guidelines for practicing engineers.

Reitherman: Here's another name on your list, the late Roger Scholl.

Bertero: We worked on some projects together. I was involved in peer review of a design project he had with Professor Bob Hanson for a bank building. He was a solid engineer. He did a lot of excellent work for EERI. I worked on one of the EERI studies on the 1985 Michoacan earthquake, studies done in collaboration with Mexican colleagues.

Reitherman: Another name on your list of engineers with whom you have worked is J. P. Singh, but I'm not sure exactly what kind of engineer he is.

Bertero: In his geotechnical engineering work he is a combination of a seismologist and an engineer—he has some good ideas concerning both disciplines. We worked very well together in some cases, and in other cases, I think there was some disruption when his consulting firm changed. There was one project we both worked on before the 1989 Loma Prieta earthquake. It was a doughnut-shaped hotel near the airport.

Reitherman: The Hyatt Hotel, just south of the San Francisco airport?

Bertero: Yes. As I commented earlier in our conversation, the design was not adequate to take a moderate or big earthquake.

Reitherman: Your conclusion was made prior to the earthquake?

Bertero: Yes. During the design phase, J. P. Singh was involved with the ground motion study and the design of the foundation, and he said he had to know also the response of the building, which is logical. I looked at the building they had designed, and I said I cannot approve this design. What I wanted with the foundation would have cost about $100,000 or $150,000 more. They did not have a sufficient number of piles. The owner decided not to spend the money. When the earthquake occurred, the hotel was damaged. The economic loss was much greater than the cost would have been to prevent the damage.

I recall Bob Preece as the owner of an experimental laboratory, Testing Engineering. He contributed considerably in the reinforced concrete workshop I organized in 1977 and also to the study of the 1985 Mexico earthquake.

Tom Tobin is someone I have known who is an expert on public policy. Tom was for many years the executive director of the California Seismic Safety Commission, and has a good understanding of the policy issues. I worked with Tom on the issue of what to do about the earthquake problems of the old City Hall in Berkeley. That is the kind of issue where you have to balance many concerns.

I have known and have worked with some of the structural engineers at Wiss, Janney, Elstner since Boris Bresler left the Berkeley faculty and became an engineer there and organized the structural branch of the office. The firm does a lot of special engineering studies, forensic work, and so on. Eduardo Fierro and Cynthia Perry are former students of mine who worked there for many years and are excellent structural engineers. They recently started their own firm. My name is associated with the firm as an advisor, but I told them: I do not want to do any forensic work. I have had bad experiences with the lawyers and the courts, and I do not need that in my life.

Besides the bad experience with the Loma Prieta earthquake legal case I mentioned earlier, I got involved with the big legal case about the Royal Palm Hotel that was damaged in the Guam earthquake in 1993. Many engineers—and even more lawyers!—got involved in that case. The technical question was whether the partial collapse of a new building was because of a short-column configuration or the detailing of the beam-column connection. But the legal questions all had to do with who was filing a lawsuit against whom and who would be blamed for the damage and have to pay for it. The case went on for a long time.

I have gotten to know Ronald Mayes talking with him about base isolation, and he gave some lectures on that subject to my graduate students. He also helped organize one of the first SEAOC annual meetings held outside the USA, in Mexico.

P. Crosby and his brother designed the Emergency Operations Center for the city of Berkeley, which I reviewed on behalf of the city and found they had done an excellent job. They also designed a building at San Francisco State University, which used for the first time Bauman's mesh. That is a product that uses prefabricated welded cages for concrete confinement. I did some testing on that. There was a lawsuit connected with that building, between the contractor and the university. So once again, I was in court. I call it being an "accidental witness."

Lee Benuska is a name I recall from long ago. He was an early student of mine. I worked with him a lot, particularly regarding the use of expansive concrete, when he joined a company interested in its application on construction projects.

Reitherman: At the CUREE-Caltech Symposium in Honor of Wilfred Iwan, held in 2006, Lee told the story of going to the 1964 Alaska earthquake and looking at the Four Seasons Building. His joke was that it should have been called the Three Seasons Building, because that's as long as that new building lasted. Fortunately, no one had yet moved in when that almost-finished six-story building completely collapsed. I think that until the complete collapse of some large parking garages in the 1994 Northridge earthquake, the Four Seasons was the largest building to totally collapse in an earthquake in the United States.

Bertero: The Four Seasons Building was a very economical building to construct, but its structural system, particularly the detailing of the reinforcement, was not adequate. The central core of reinforced concrete walls was built first, then the lift-slab floors, which were post-tensioned, were put in place. This disaster illustrates the problem that can be created by the use of a structural system that relies on a core without significant frame lateral resistance around the perimeter, and connections of thin, prestressed slabs to the core walls, and by poor detailing of the reinforcement. Unfortunately, new tall buildings are at present being designed and constructed using similar basic structural systems. We are not learning our lessons.

Reitherman: Dan Shapiro is one of the practitioners you note as having known. The current initials of the San Francisco structural engineering firm he was a partner in, SOHA, came from the partners' names, Shapiro, Okino, and Hom, and I think the "A" is for associates.

Bertero: Dan Shapiro has hired some of my former students. I remember when he was working on the ATC 33 seismic retrofit project, as the leader of the management team. ASCE was also involved in some way in that effort, and that's where I was doing some review work. Dan is not only a very capable structural engineer but also a true gentleman and a pleasure to work with.

Reitherman: Also on your list of practicing engineers you have known is Chuck Thiel, who is also known for his work within the federal government.

Bertero: Chuck is a very smart person. Chuck, of course, has been involved in many things in the earthquake field, from his days at NSF when NEHRP was just starting. Before, and especially after, the Northridge earthquake he was very busy doing consulting work for the state university system. We have had many good discussions about earthquake engineering, particularly after the Northridge earthquake.

Reitherman: I recall being in the facilities engineering building on the California State University of Northridge campus after the Northridge earthquake in 1994, looking at drawings while I was doing EERI reconnaissance survey work there. Chuck walked in wearing a hard hat, carrying the biggest crow bar I've ever seen, grabbed a big roll of drawings without hardly stopping, and strode out. It was another side to the Chuck Thiel most people know as the intelligent, articulate type wearing coat and tie.

Bertero: Yes. He has had a varied career in earthquake engineering, from his early years at the National Science Foundation as a manager of the earthquake program, being instrumental in the formation of NEHRP, later at FEMA, and then being a consultant.

I also knew Jack Meehan, who was in the state architect's office for many years. We did some testing in the 1960s at Berkeley for that office. We were looking at wood panels that were used in school construction. He was very thorough in making sure school construction was safe. He went out looking at earthquakes to see what he could learn that applied to California schools.

Early Use of Computers in Consulting Engineering

Bertero: I mentioned that I knew several young professors at MIT who established prominent consulting engineering firms: Howard Simpson, Werner Gumpertz, and Frank Heger; Robert Hansen, John Biggs, and Myle Holley, Jr.; William LeMessurier. Later, after I was at Berkeley, I had some of my own experience doing consulting engineering work.

The first consulting job I had after I arrived at Berkeley was to work on a tall, guyed, steel truss tower on the island of Guam to be used for communications. There were tremendous wind design loads for Guam, and the structure was a complicated one. I had done some research on that type of problem at MIT. The truss tower and its guy cables are designed to remain elastic, but as it deflects under wind it does so in a nonlinear way. It was a difficult problem I had worked on at MIT for a consulting project Professors Holley and Biggs had had, so I knew something about this problem, and I was sought out by the consulting firm, McCreary-Koretsky Engineers, which had the project for the U.S. Navy.

We developed a computer program. All those punch cards! But we needed a big computer to do all those computations. After six o'clock in the evening, I had to travel to San Francisco where the consulting firm could use a big computer that was used in the daytime for the Blue Cross and Blue Shield health insurance organization. That was in 1959.

Reitherman: In 1959, you couldn't use a big computer on the University of California campus, and the consulting firm didn't have a big computer or access to one?

Bertero: No, we had to use the one that Blue Cross and Blue Shield had in San Francisco, and we had to work all those hours in the night. Preparing all those IBM punch cards, and if any little thing went wrong, oh my! It was hours and hours of work all over again.

Reitherman: Are there any negatives to the extensive reliance on computers in structural engineering today?

Bertero: The big problem was some years ago, when engineering students had to take many hours of classes to learn programming and how to use programs, which took away from other classes that were important for the education of an engineer. Maybe we needed one more year in undergraduate education, but the advances in hardware and software have reduced the problem, because students can devote their minds to the subject matter. It's the same with the practicing engineers. Without the development of the computer, practicing engineers could never use the analysis methods they use today; it would take too long. It is partly a development of the computer hardware, also the sophistication of the software, and also the ease with which the engineer can use that hardware and software. Today, you don't have to take computer programming classes and be a clerk handling all of those punch cards. It is a big improvement.

However, to answer your question, yes there are some negative aspects to the reliance on computers that we should be concerned about. It is unfortunate that there has been a trend among the young practicing engineers who are conducting structural analysis, design, and detailing using computers to think that the computer automatically provides reliability. This problem is aggravated by economic pressures to quickly use the computer to select a system, design it, and have it built at the lowest possible cost, without conceptualizing how the whole building will actually perform. You can check off each little requirement of the building code one by one and miss some larger issues. Engineers still need to understand the actual physical behavior of their buildings, the mechanical and dynamic behavior, when the total system undergoes unusual but expectable hazard demands during its life. By total system I mean the whole building, not just the bare structure.

Chapter 9

Earthquake Engineering in Latin America

In general, in Latin American universities, in the civil engineering department there are very few full-time professors. They must do their outside consulting work to make a full-time income.

Reitherman: From your curriculum vitae, I see that you have received a number of awards from Latin American engineering societies and universities.

Bertero: The reasons for these awards are not only because of the research that I have done in the USA, but also because I have given lectures and taught many short courses, even though most of these countries do not have sufficient funds to pay you to teach. It is for my teaching of short courses on earthquake engineering and in the discussions of the basic seismic engineering research that is needed in those countries that I have received awards.

Argentina

Bertero: To the best of my knowledge, there are at present only about three or four universities in Argentina that are teaching earthquake engineering, including the universities at Córdoba, Mendoza, and San Juan. From the five universities that existed in the country when I was in college, there are now hundreds, but earthquake engineering is not widely taught. The other universities don't teach the subject, but they produce many engineers who will be designing structures that should be earthquake-resistant. When you receive an engineering diploma from a university in Argentina, it is the same as a license to practice engineering anywhere in Argentina. The biggest and best known university in Argentina is in Buenos Aires, which does not teach earthquake engineering, but it produces many practicing engineers who later could be involved in the design and/or construction of buildings as well as other facilities located in zones that can be subjected to significant earthquake ground motions. Therefore, they need to learn earthquake engineering when they are university students.

In general, in Latin American universities, in the civil engineering department there are very few full-time professors. They must do their outside consulting work to make a full-time income. The engineering professor is the professor of one or two courses, which are taught again and again. Many of the best practicing engineers are professors. In the United States, most professors receive a full-time income. Sometimes that situation in Latin America is a benefit, because there is less separation between the profession and the teaching of it.

Reitherman: It is obvious from the honors on your résumé that Argentina does not begrudge the fact that one of its own went off to America to be a famous engineer. I know you won't brag about these honors, so let me read off some that have been conferred on you by your Argentine compatriots: elected to the Academy of Science of Argentina, 1971; Honorary Professor and Honorary Research Advisor, University of Rosario, 1983; Ingeniero Enrique Butty Award, 1988, from the Argentine Academy of Engineering; elected to the Argentine Academy of Engineering (Academia Nacional de Ingenieriá), 1989; Honorary Professor, Universidad de Buenos Aires, 1991; Honorary Member, Structural Engineers Association of Argentina, 1992; Honorary Professor of the Universidad Tecnologica de Mendoza, 1992; Diploma Mérito from the Konex Foundation as one of the five most outstanding Argentine civil engineers of the decade, 1993; *Doctorado Honoris Causa en Ingenieriá*, University of Cuyo, Mendoza; *Ciudadano Ilustre* (Distinguished Citizen), City of Esperanza, 2006; Honorary Academician of the Argentine National Academy of Engineering, 2006.

Ecuador

Reitherman: The engineers and universities in Ecuador seem to have appreciated your efforts as well. Your curriculum vitae shows that in 1979, the University of Guayaquil made you an honorary professor. Also in 1979, the Structural Engineering Association of the Guayas made you an Honorary Member. Just recently, August 13-17, 2007, the Third Seismic Engineering Conference of the Seismic Society of Ecuador was named after you.

Bertero: I have given short courses in Quito, the old capital, and also in the more modern city of Guayaquil. As I mentioned, I appreciate these honors, but regard them more as expressions of gratitude for the teaching I have done in Latin America than as awards.

Colombia

Bertero: I have lectured at the National University in Bogotá, the *Universidad Nacional de Colombia*. The key leader in the field of earthquake engineering has been Professor Alberto Sarria, who in 1990 published an excellent book entitled *Ingeniera Sismica*. Professor Sarria started to write this book in 1970 as notes of a course on seismic engineering he offered at the Department of Civil Engineering at the *Universidad de los Andes* in Bogotá. I met Professor Sarria, as well as other engineers from Colombia, in a seminar given in Spanish on design for lateral loads offered by ACI and held in Miami in 1985 when I presented a summary called "Seismic Behavior of Reinforced Concrete Structures." At present, there is an excellent group of professors and professionals working in the earthquake engineering field in Colombia, among them Omar D. Cardona and Luis E. Garcia. Garcia has been a professor at the *Universidad de los Andes*, where he got his undergraduate degree before getting his Master of Science from the University of Illinois, and is now head of a consulting engineering firm. He was voted vice-president of ACI in 2006. The Colombians have developed excellent seismic code provisions, especially for dealing with the effects of irregularities in structures.

Mexico

Reitherman: When did you first meet Emilio Rosenblueth?

Bertero: It was 1964, and it happened to be at a meeting in Peru, not in Mexico. Professor Rosenblueth and I became very close friends. Emilio was smart, very smart. You could call him a scientist. Emilio was the one who established a very good group at the National University in Mexico City, *Universidad Nacional Autónoma de México* (UNAM). Luis Esteva, for example, worked with Rosenblueth and got his PhD there after he got his master of science from MIT.

Rosenblueth was so far above the ordinary practicing engineer that I think sometimes he could not understand what they needed from him and the university. After the 1985 earthquake disaster in Mexico City, he called me up and we talked. He said, "Vitelmo, now I understand why you university people in California have so many activities with the practicing engineers, why your Structural Engineers Association of California created a new organization, the Applied Technology Council, just to take existing research and apply it to develop guidelines that the design engineer can use." In my work, I have tried to see things through the eyes of practicing engineers and understand their problems so I could teach them about earthquake engineering. This is a different kind of activity than preparing your PhD students.

Emilio had a large influence on the early work of ATC. When the large group of us was working on ATC 3-06, from 1972 to 1978, he was a member of the ground motion committee.

Rosenblueth had a big effect on many people in the field. Anil Chopra here on the Berkeley faculty became close with Emilio. You know, Professor Chopra did his PhD here with Professor Clough, then he was a professor for a time at the University of Minnesota. He wrote to say how cold it was there! We brought him back to Berkeley and he has been on the faculty since.

Reitherman: Was Rosenblueth involved in the design of the Torre Latinoamericana in Mexico City? I think he had finished his PhD at the University of Illinois by the time that building was in design.

Bertero: No, I don't think so. The engineers for the building were the Zeevaert brothers, Adolfo and Leonardo, and Nathan Newmark was the seismic consultant.

Reitherman: For many years, that 44-story building was the tallest structure in a high-seismic zone, much taller for example than anything in California or Japan. It is known for its good performance in both the 1957 and 1985 earthquakes that affected Mexico City.

Bertero: Rosenblueth was a very, very smart man. The book he wrote, *Fundamentals of Earthquake Engineering*,[44] with Nathan Newmark, who was his professor at Illinois, is still one of the best books available.

Reitherman: That's high praise, considering that the more up-to-date anthology reference work that you co-edited with Yousef Bozorgnia is so highly regarded.[45] I see from the course

outline you handed out for your graduate course on Structural Design for Dynamic Loads—the copy I've seen is from 1974 — that in addition to your own notes you gave to students, and a long list of about twenty references, the one textbook reference you recommended was that book by Newmark and Rosenblueth. Actually, it's hard to tell how many references your students were encouraged to consult because I see that "one" reference was "Earthquake Engineering Research Center Reports." In his EERI oral history, Joe Penzien added up the EERC reports published each year, and even by 1974 there had been ninty-four published.[46] And the last reference on your reading list was "selected papers published in technical journals and proceedings of symposiums and reports on recen t destructive earthquakes." None of your former students has ever complained that you didn't give them enough work!

Bertero: To produce the book *From Engineering Seismology to Performance-Based Engineering* required me to initially review the chapters submitted by our co-authors about five times and give them guidance. There were some problems working with the publisher, and I had to go over the chapters another three times as the book was being finalized.

I gave short courses in Mexico City at UNAM, the national university, and also at UAM, which is the *Universidad Autonomous Metro*, or the Metropolitan University. Now there is some competition for UNAM because of

44. Nathan M. Newmark and Emilio Rosenblueth, *Fundamentals of Earthquake Engineering*. Prentice-Hall, Englewood Cliffs, New Jersey, 1971.

45. Vitelmo Bertero and Yousef Bozorgnia, editors, *From Engineering Seismology to Performance-Based Engineering*. Boca Raton, Florida: CRC Press, 2004.

46. *Connections: The EERI Oral History Series— Joseph Penzien*, Stanley Scott and Robert Reitherman interviewers. EERI, Oakland, California, 2004, p. 42.

UAM. But prior to the establishment of UAM, there was really only UNAM. I was in Mexico City for the twentieth anniversary of the 1985 earthquake. I was surprised that there had been so little communication of the knowledge of the universities to the practicing engineers, especially outside of Mexico City, and the engineers were complaining about that. I have also given lectures several times at different universities in Guadalajara.

Reitherman: Argentina, of course, has honored you many times, but it's fair to say that you have built up widespread respect in Mexico also. Your curriculum vitae shows that in 1983 you received the Commendation of the Mexican Society of Seismic Engineering, and later that year the Diploma of Recognition for earthquake engineering achievements from the Mexican Seismic Engineering Society. In 1986 UNAM conferred on you the Extraordinary Chair of the College of Engineering; in 2002 you were honored as the Honorary Member of the Mexican Seismic Engineering Society.

Bertero: After the 1985 Mexico earthquake, I took on the responsibility for EERI of editing the report on the research in the U.S., Mexico, and Japan on that earthquake.[47] I considered that the lessons learned from the studies on the Mexico earthquake did not get the attention that they deserved. The main reasons for this were, first, before we finished the discussion of the studies, the 1988 Armenia earthquake occurred. Secondly, when we finished

the report and EERI published it, the 1989 Loma Prieta earthquake occurred. Anytime a significant earthquake occurs, researchers go immediately to the field, where the earthquake ground motions have induced damage—in other words, where the real experiment has been conducted—and begin investigating the reasons for the damage. However, in the case of the Mexico earthquake, the Armenia and Loma Prieta earthquakes distracted attention from what should have been learned from it. The Japanese, who had done a large amount of research on the Mexico earthquake, did not participate in the final discussion of the report—they, too, had their attention diverted.

A number of good recommendations came from research on the 1985 earthquake in Mexico City. One of the big problems was the fact that a large earthquake 300 kilometers away from Mexico City could cause such damage. That can happen in other places in the world also.

Another problem addressed was that of general adjacency. Mexico City had so many buildings close together. So there was pounding. Even without pounding, when a building collapsed it could fall on its neighbor across narrow streets.

Reitherman: Like the toppling of one of the towers in the Pino Suarez complex onto an adjacent building?

Bertero: Yes, that's a very dramatic example—a twenty-one-story steel building collapsing on top of a fourteen-story one, with debris extending across the boulevard into the next block.

---

47. Vitelmo Bertero, editor, *Lessons Learned From the 1985 Mexico Earthquake*. Earthquake Engineering Research Institute, Oakland, California, 1989.

Peru

Reitherman: What about Peru?

Bertero: Because Peruvians are very proud of a local drink of theirs, I will tell you a little story about it that I remember from 1966, when there was an engineering conference in Peru. I was with a professor from the USA. He had a little too much of the local alcoholic drink called *pisco*. It doesn't taste strong—but it is actually very strong. It can be 90 proof. He did not feel well afterward for two days.

Reitherman: Those mild-but-strong drinks are like rattlesnakes without rattles.

Bertero: Pisco can sneak up on you.

The key leader in the earthquake engineering field in Peru has been Julio Kuroiwa. He was one of the first students at the International Institute of Seismology and Earthquake Engineering that the Japanese have been running.

The big seismic problem in Peru is informal construction—self-built dwellings and shops without any engineering or code compliance. Some people call it vernacular architecture, but I call it informal construction. Just pile up whatever local materials are at hand without any consideration of the earthquakes or other hazards the buildings will face. Dr. Marcial Blondet, who used to be here at Berkeley and now is at the Catholic University of Peru, *Pontificia Universidad Católica del Perú*, has done some very good work on how to reinforce adobe buildings, and in general how to make masonry construction earthquake resistant. But the big problem is how to train the people to use the techniques.

I have taught short courses at the Catholic University and at the National University in Lima, and also in the university in Chiclayo near the border of Peru and Ecuador, an old city with Incan ruins.

Chile

Bertero: Since I mentioned a story about pisco in Peru, I should also mention one that happened in Chile. The Peruvians and the Chileans both are very proud of their pisco, the strong type of brandy, and they have heated debates about who has the better quality. It is an international controversy between the Peruvians and Chileans as to which country has the better pisco. In Peru, a cocktail called a Pisco Sour is made by adding lime juice, some sweet syrup, a few drops of bitters, and egg whites to the pisco. In Chile, the Pisco Sour has no egg whites or bitters, and they use lemon instead of lime juice. It is a very refreshing drink on a hot day. I visited Chile in 1969 for the Fourth World Conference on Earthquake Engineering. I remember I was sitting in the hotel near the reception area and met others who were arriving from the USA. I was having a small pisco—not much pisco and lots of ice.

As I sipped my diluted pisco drink, several colleagues of mine who arrived had martinis. The others asked me about my pisco drink and insisted on ordering large ones. They really enjoyed that drink. Then we left to go to a very nice hotel on the mountainside for dinner. My friends ordered more pisco drinks, and, in other words, they became very happy. The waiter talked to me in Spanish and asked if I could get my friends to be a little quieter, because they had started singing.

One of the consulting engineers at our table at this point had gone out, fast asleep from his drinking, his head on the table. We got him in the taxi. Back at the hotel, another professor from Berkeley and I helped him up to his room, carrying him and putting him on his bed. The next day I saw him and he asked me how he had gotten home.

Reitherman: Yet another example of the way you and your U.C. Berkeley colleagues have supported the practicing engineer! Concerning practitioners in Chile, is the connection between the academic and the practicing engineer a close one?

Bertero: Yes, partly because there are few really full-time civil engineering faculty positions in the universities, the academics have to practice to make a living, the same as in the rest of Latin America. That tends to make it hard on academic research, but it helps the connection between the university and practice. H. Bertling, a professor of applied seismology at the University of Chile, in Santiago, wrote a very practically oriented paper for the 1956 World Conference in which he summarized the history of construction styles in Chile along with the development of seismic codes.[48]

In 1960 when the big earthquake occurred in Chile that caused such devastation in Concepción and Valdivia, I started to become acquainted with the work in the field of earthquake engineering of Professor Rodrigo

Flores, who then was the one in charge of the Fourth World Conference on Earthquake Engineering, which was held in Santiago in 1969. Have you heard of Professor Flores?

Reitherman: Yes. He must be one of the best known Chilean engineers and regarded as one of the founders of the modern era of the field there.

Bertero: Yes, we know him for that. He was not only an excellent professor and professional engineer, but the best chess player in Chile. His first national championship was back in 1931, and he continued to win national and international championships for decades.

The huge earthquake in Chile—actually it was several large earthquakes—in 1960, with its main shock having a magnitude over 9, was really an aftershock, since there was a sizable earthquake the day before. There were a number of earthquakes in that series, with that one huge event that has held the record as the largest magnitude earthquake ever recorded. That earthquake caused a tsunami that killed people in Hawaii and led to a great advance in tsunami warnings in the Pacific.

About 1965, the Rector, which is the same as Chancellor, of the Catholic University of Chile in Santiago made a visit to U.C. Berkeley. One day I received a telephone call from the Chancellor's office. They explained who was visiting, and that he spoke mainly Spanish. So I met with him. He said he did not want to go out to lunch, he did not want a nice tour. He wanted to find out how to modernize the structural engineering division of the civil engineering department of his university. He wanted to send the university's

---

48. H. Bertling, "Development of Earthquake-Proof Construction in Chile," *Proceedings of the World Conference on Earthquake Engineering.* Earthquake Engineering Research Institute, Oakland, California, 1952.

best students, with a scholarship, to Berkeley to study, and to have a Berkeley faculty member go to Chile in the summer to teach a short course. I said that the budget at Berkeley would not allow a faculty person to be paid the expenses of going to Chile to teach, but I knew that some other universities in the U.S. had arranged some other funding. Bob Whitman at MIT was doing some research in Venezuela. There was some sort of fund that supported the cost of MIT professors going to Latin America.

Reitherman: The Ford Foundation?

Bertero: Yes, that's it. And the same foundation allowed C. Martin Duke and UCLA professors to go to the National University in Santiago. But the UCLA professors were dedicated to that separate university. So I suggested to the Rector of the Catholic University of Chile that he should go to the Ford Foundation and simply tell them what he was trying to do. If the Berkeley professor's expenses to go to Chile could be taken care of by that other funding, the plan might work. I called the Chancellor's office at Berkeley to see if this was a good idea, and I talked to him. My Chancellor agreed it was a good plan, and so that is what happened. The Catholic University chancellor went to the Ford Foundation, and they agreed to provide some funding.

After that, there were several very good Chilean students who came to Berkeley for their PhDs, and they returned to Chile to join the faculty of the Catholic University. One of them was a student of Professor Penzien. His name was Patricio Ruiz, and he became the one who reorganized the structural engineer-

ing program at his university in Santiago. Later came Jorge Vasquez who had Professor Popov and myself for his PhD supervisors. Then Pedro Hidalgo came to Berkeley and got his PhD under the supervision of Professor Clough, and later Ernesto Cruz who got his PhD under the supervision of Professor Chopra, and the last one that I remember was Juan de la Llera.

The first Berkeley professor to go there was Professor Clough. Professor Clough asked me how courses were taught in Chile. I told him that if he taught after lunch, he had to know that everyone was used to taking two hours for lunch, having the big meal of the day, and drinking wine or something with it. Without a nap afterward, it would be hard for students to concentrate, and hard for teachers to teach.

When he came back, he said, "Vitelmo, I know what you mean. I would go to lunch with them, and in my afternoon lecture it was very, very difficult."

Reitherman: Who are the Chileans with whom you have worked?

Bertero: One has been Rodolfo Saragoni, who is at the National University in Santiago. Saragoni was educated in Los Angeles, and a colleague of his, Arellano Sarrazini, was educated at MIT for his master's and doctorate. I already mentioned Rodrigo Flores, who can be called the father of the field of modern earthquake engineering in Chile. I have taught short courses at the National University of Chile, Santa Maria University in Valparaiso, and Catholic University and *Universidad de los Andes* in Santiago.

Reitherman: How do you explain the fact that the structural engineer in Chile has been able to get the architect to allocate such a large quantity of structural walls in their building plans? In the 1985 earthquake, for example, at Viña del Mar, the performance of the mid-rise concrete buildings was unusually good, as documented by Sharon Wood,[49] and that was attributed to the large number of structural walls in their layout.

Bertero: The architects respect their engineers more there. They have some good architects, but they still respect their engineers. They know they have a large earthquake problem and take it seriously.

Reitherman: Do the architecture students there take more structural engineering than in the U.S.?

Bertero: Yes, they do. I've said how here at Berkeley the structural side to the architecture students' education has been reduced. In many other countries, architects are better educated in the technical aspects of building construction.

Reitherman: Because countries in Latin America have differing building codes, is it difficult to teach a subject like earthquake-resistant reinforced concrete construction?

Bertero: In Chile, the code provisions now are based on the American approach. For example, the reinforced concrete code follows the ACI approach. Many of the other countries have reinforced concrete design more based on what the Europeans do. The design provisions for the materials are not that different, but the seismic regulations differ, as do the way the engineers practice and especially the way the buildings are constructed. Chile has good construction quality for earthquake resistance, as compared to Argentina, for example, except in the west of Argentina around Mendoza and San Juan where there have been frequent earthquake disasters. There have been some multistory buildings that have collapsed in Buenos Aires, even without an earthquake. That is due to inadequate construction, not a lack of engineering knowledge on how to make a building stand up under the predictable gravity loads.

Reitherman: If earthquake resistance is built into the structure, does that tend to automatically make the structure more reliable in general, such as in safely resisting everyday gravity loads?

Bertero: Yes. Designing for earthquake resistance tends to make the entire structure more reliable. You tie the structural elements together. You worry about ductility. You work hard to obtain good construction quality. You worry about the uncertainties.

Costa Rica

Bertero: The one who led the development of the field in Costa Rica was Franz Sauter. He has passed away. He was a very good engineer and a good writer. Costa Rica is one of the best countries in Latin America for earthquake

---

49. Sharon Wood, "Performance of Reinforced Concrete Buildings in the 1985 Chile Earthquake: Implications for the Design of Structural Walls," *Earthquake Spectra.* Vol. 7, no. 4, November 1991. Earthquake Engineering Research Institute, Oakland, California.

engineering, though in San José they have some vulnerable buildings. Several times I went to San José and gave lectures there. Jorge Gutierrez, who got his PhD at Berkeley, is a leader of earthquake engineering in Costa Rica now.

The best educated engineers in Costa Rica, along with those from Mexico and Chile, are on a high level, I would say. Again, a society that has good engineering but does not apply that knowledge in the codes, the practice, the construction, will have big earthquake problems. The government needs to assist that process.

Reitherman: From your curriculum vitae, I see that in 1994 you received a diploma award from the Earthquake Engineers Association of Costa Rica. What is that organization like?

Bertero: This has been a very progressive organization. I believe that Franz Sauter played an important role in this association, improving the code provisions for seismic resistant design and construction. There is a relatively young group of academic and professional engineers that knows how to adapt available information for the formulation of simple but reliable seismic design that meets the needs of the country.

Venezuela

Bertero: There were five graduate students at Berkeley whom I had for their master's work: Irragori Montero, Juan Pereda, A. Cova, I. Uzcategui, and L. J. Alonso. They were back in Venezuela at the time of the 1967 Caracas earthquake. Have you been to Caracas?

Reitherman: Yes. You begin at sea level on the coast where the airport is, then go up over the first range of mountains to find that big city filling the valley.

Bertero: That valley, with its soft soil, was really one of the big aspects of that earthquake. Professor Harry Bolton Seed had already begun to study the influence of soils on shaking after the 1964 Alaska earthquake, and then he had a perfect case to study in Caracas. That topic advanced very much because of Seed's work on that earthquake.

When I have been back to Caracas, oh my! I go there now and they didn't learn the lesson. They are still building the kinds of buildings that collapsed in 1967.

I remember studying the Macuto Sheraton Hotel, with its huge concrete columns, which were shattered because of a soft story at that level. The Charaima Building was one that I studied in detail with Joe Penzien and Steve Mahin.

Reitherman: As of the 1967 earthquake and for a few years after, is it true that the Uniform Building Code in use in California did not yet include modern ductility requirements for reinforced concrete construction?

Bertero: That's right. It took several years. The book by Blume, Newmark, and Corning had already been published, which called for ductility, but the code and practice had not yet adapted to the new information. That was the other main impact of the 1967 Caracas earthquake on earthquake engineering, along with the lessons on the effect of local soils on shaking, and the vulnerability of buildings with their wall system discontinued at the ground story for architectural reasons.

Reitherman: One of your former master's students, Eduardo Fierro, recently co-authored an article[50] that presents a pessimistic view of how vulnerable current construction in Venezuela is, and how little the lessons of the 1967 earthquake there have been absorbed. Age-old problems are being repeated there, according to the survey, such as the soft ground story. Masonry infill walls in concrete frames are provided above that level, but those walls are dispensed with just where they are structurally most needed, because of either ground-story parking or wide-open lobbies and other entry-level public occupancies.

Bertero: In the hills around Caracas, some large new houses have a soft ground story, and I saw one that also cantilevered out of the hill. That design causes two big earthquake problems—the vertical configuration problem of the soft story, and the problem, in plan, of torsion. Another of my former students, Teresa Guevara-Perez, who received her PhD in architecture at Berkeley, sent me some photos of these incredible designs that defy basic seismic design concepts. The fantasies of the architects are just unbelievable. Teresa recently sent to me a copy of the final draft of a voluminous book in Spanish she has written called *Introducción al Estudio Sistemático de las Configuraciones Arquitectónicas Modernas en Zonas Sísmicas*, which translates in English to *Introduction to the Systematic Study of*

Modern Architecture Configurations in Seismic Zones. Part of it will be published in Spain by the Editorial Gustavo Gili. It is an excellent book. The information collected and the results obtained in the studies conducted are very well discussed and illustrated by excellent figures, sketches, and photos. It would be of great interest not only to architects and city planners, but to all those involved in the implementation of the practice of modern seismic engineering.

Teresa Guevara-Perez got her start on this topic while doing her PhD in the architecture department at Berkeley, being helped by Henry Lagorio there and also Christopher Arnold. She makes reference to the book you and Chris Arnold wrote on building configurations being especially influential. It's a pioneer in its field.[51]

Reitherman: I think the fact that it has been translated into Spanish, also Italian and Russian, has helped its appeal. It was the brainchild of Chris Arnold and Eric Elsesser, the structural engineering consultant on the NSF-funded research project that led to the book.

Bertero: Since the early 1960s, I have visited Venezuela several times to give short courses or lectures at the *Universidad Central de Venezuela* at Caracas, also the university at Barquicemento. The name of the city is related to cement, and it is a big industrial center for that. I have also taught short courses at the *Universidad de los Andes* in Mérida, up in mountains, a beautiful old traditional city.

50. Gary Searer and Eduardo Fierro, "Criticism of Current Seismic Design and Construction Practice in Venezuela: A Bleak Perspective," *Earthquake Spectra*. Vol. 20, no. 4, November, 2004. Earthquake Engineering Research Institute, Oakland, California.

51. Christopher Arnold and Robert Reitherman, *Building Configuration and Seismic Design*. John Wiley and Sons, New York, New York, 1982.

Reitherman: Having been to Mérida, let me ask you a leading question: What kind of transportation did you take to get there?

Bertero: Oh my! It was an airline flight, but not a typical one. You look out the windows as you fly up the valley with mountains on both sides. It's an adventure.

Reitherman: And you land going noticeably uphill on the runway. It seems odd that a runway wouldn't be level, but you realize it's a good thing, so that the plane, a regular jet airliner, not a small propeller-driven plane, can stop, barely, before it gets to the fence at the end of the single runway. Apparently the crews that fly there receive extra pay for the hazard duty.

I see from your curriculum vitae that you not only were there in 1977, and were given a Distinguished Guest honor from the city and an honorary diploma from the university, but you were brave enough to return in 1987, when the University of the Andes gave you a Bicentennial Medal, and again in 1993 when the University awarded you an honorary doctorate, the *Doctorado Honoris Causa en Ingeniería*. Then later that year you returned to receive the title of Honorary President of the Latin-American Society of Earthquake Engineers, along with a medal from the state government of Mérida. You also received the Institutional Medal of the *Universidad Centro Occiddental Lisandro Alvarado Barquisimento*, 1993.

You are still the honorary president of the Ibero-American Society of Earthquake Engineers, the *Asociación Iberoamericana de Ingeniería Sísmica* (AIIS). How do you compare it with an organization like EERI?

Bertero: Yes, still I am the honorary president of the Association. It was created in 1992 during the Tenth World Conference on Earthquake Engineering in Madrid, thanks to the efforts of Professor Rafael Blázquez, who was the head of the 10WCEE steering committee. This was done because of the necessity to coordinate the efforts in earthquake engineering and related fields. The first organizing meeting was on the occasion of the Eighth Latin American Seminar on Earthquake Engineering that was held in Mérida in 1993. Every four years we have such a seminar. There is a problem regarding the funds necessary for organizing such conferences. There are about ten countries that are members of AIIS, but at present there is not agreement regarding who is in charge of making decisions. Thus, AIIS cannot be compared with EERI, which has been well-organized for so many years. It is my hope that this situation of the AIIS can be solved.

Dominican Republic

Bertero: I started first in the 1970s giving some lectures in Santo Domingo about the activities of ACI regarding earthquake resistant design of reinforced concrete structures, particularly building structures. Then I participated in a conference organized in Santiago de los Caballeros. Since 2002, Eduardo Fierro and I have offered short courses on earthquake engineering there. There is a very capable group of engineers in this field in the Dominican Republic.

Reitherman: Were most of these short courses that you taught in Latin America on reinforced concrete seismic design?

Bertero: Yes, because steel is usually too expensive. But in the Dominican Republic, about the year 2000, the construction industry started to build more in steel; not using rolled steel structural sections, but rather light-gauge cold-formed light steel. They are making buildings four or five stories tall with the structural system being designed and constructed using these light-gauge members.

Reitherman: Are they framing up stud shear walls with that kind of material?

Bertero: No, they are used to make frames.

Reitherman: But you can't weld those thin sections, can you? How do you make the moment-resistant connections?

Bertero: That is the problem. Although I have seen frames up to two stories in which they weld the beams to the columns, I hope that they are not doing this type of construc-

tion for taller buildings. It reminds me of when I was practicing in Argentina in the late 1940s and reinforcing steel was introduced with a higher strength because it had already been strained, by twisting, but at the expense of ductility. I don't think they are getting ductility in this new multistory light steel frame construction in the Dominican Republic.

Reitherman: Yet another mark of how well known and respected you are in Latin America: you received an award from the Earthquake Engineering Society of the Dominican Republic in 2006.

Bertero: As I have said, these awards are in recognition of my contribution as a teacher in these countries, giving short courses. I enjoy teaching. I do not think we should give the impression that I have received all these honors for particular scientific contributions in these countries.

Chapter 10

Major Unsolved Problems in Earthquake Engineering

The engineer has to visualize what the numbers that are obtained from the numerical analysis physically mean.

Understanding Rheology

Bertero:　Rheology is the science of how any material deforms as it is stressed. In some fields, rheology is related to the flow of fluids. For the structural engineer, rheology is the science of how the solid material, such as concrete or steel, deforms. The actual behavior of a structure in an earthquake is a very, very complex phenomenon. Engineers want simplification so they can accomplish their projects. We need simplification. But you have to really understand why you can rely on a simplification. You must feel the physical behavior.

For example, consider a reinforced concrete structural member in a structure subjected to an earthquake. Even relatively small tensile strain in the concrete during an earthquake means cracking—that is, there is some damage. The earthquake ground motions continue and you get a crack in the other direction. The earthquake continues and you get more cracking, and then spalling. Inelastic behavior of more than just the concrete cover could occur as well, with yielding of the reinforcing steel bars in tension and compression. The engineer has to visualize what the numbers that are obtained from the numerical analysis physically mean.

Reitherman: To give the nonengineering reader some feel for the cracking of the concrete in tension, can we say that a typical amount of strain to cause a crack in concrete is 0.001? That is, elongation of concrete by 1/1000 of its initial length. So, if a one-story-high column (say three meters) extends by three millimeters, there will be a slight crack? That is about the thickness of three or four sheets of paper. Stretch the concrete column by that small amount, and you get a crack. Correct?

Bertero: Yes, deformations that seem small to people in daily life can be significant to the structure. The designer has to imagine how the materials will feel when those deformations occur.

Today, with the use of the computer, we have lost two kinds of engineering education. One is drafting. When the structure was sketched and the members and details drawn as you thought about them, there was more thinking going on as you made those decisions. Computerized drafting is faster, but there is less thinking going on. And the other is materials testing, understanding the actual physical behavior of materials from testing them, alongside of learning the analysis. The engineer needs to

see and feel how the structure will behave. I don't think computer programs by themselves will solve the earthquake problem.

Engineers are familiar with the graph showing how steel deforms under force, as you pull on a specimen in a testing machine. Usually, you put a strain gauge on the specimen and you get one value. You put several strain gauges closely spaced, and you get more values, but some differences. Why?

With one strain measurement representing what is happening over a length of the steel bar, past the elastic limit the graph flattens out, before eventually climbing a little bit from strain hardening, and then finally there is failure. The flat portion of the curve implies that the modulus of elasticity is zero. If this is true for an entire section and length of the steel bar, there is no way that the structure can stand up. If the bar has no stiffness, it will buckle under any compressive load. But, this is not true. It is not true that 100 percent of the steel in a given length of the bar goes inelastic. When you carefully observe the tests of steel bars, for example, when you look carefully at the lines that appear in the whitewash painted on the bar, you see that only narrow bands of material went inelastic. These are called slip lines, or Lueders' lines.

But next to those signs of inelasticity, you see material that is still elastic. If you measure the strain across all those little elastic and inelastic regions, you just get an average, and it plots as an increase in deformation without increase in force. Significant advancements in knowledge about the behavior of steel members considering the true stress-strain relationships rather than the nominal stress-strain curves, was obtained from research conducted at Lehigh University in the late 1940s and in the 1950s.

All this is part of the rheology of the structural materials that engineers rely on in their design work. The engineers must understand how their materials are behaving.

Recognizing the Risk From Distant Earthquakes

Bertero: I received a telephone call yesterday [2007] from Argentina, about a tall hotel being designed for the city of Mendoza. They are using coupled shear walls. It is being designed in Buenos Aires. The connecting beams are thin, only about 25 or 30 centimeters in width. They are about a meter high and have very short spans. We know you should put diagonal bars in the beam to connect it to the walls on each side. This is from the work of Professor Tom Paulay in the 1960s.

People who have studied earthquake engineering know this. But the engineers in Buenos Aires wanted to follow that advice to use diagonal reinforcing by slanting the diagonal bars until they were practically horizontal, because there was a lot of congestion of the steel bars in those relatively thin beams. I told them, "It will not work. It simply will not work." If the designer does not understand why the steel bars should be arranged a particular way in order to achieve the intended ductility when the structure is subjected to the effects of the earthquake ground motions, then you get these sorts of shortcuts that will weaken the structure if you don't watch out. In Buenos Aires, when you mention earthquakes, they say, "That is a problem to the west in San

Juan and Mendoza, not here," and they do not think about it. There have been earthquakes in the area of San Juan and Mendoza in 1861, 1894 (which had the largest magnitude), and in 1944. The last earthquake in the San Juan area, the Caucete earthquake, was in 1977. If you don't have good university education in earthquake engineering, it will be difficult to have sufficient knowledge among the average professional earthquake engineers.

Reitherman: Another Latin American building with coupled shear walls comes to mind, the Banco de America in the 1972 Managua, Nicaragua earthquake.

Bertero: Yes, that building, designed by T. Y. Lin, was square in plan, seventeen stories in height. It had a dual structural system for resisting earthquakes. The system consists of a combination of a central core of coupled reinforced concrete walls, with a perimeter reinforced concrete frame that really acts like a tube. The central square core consists of four concrete shear wall cores for stairs and other services located at the corners of that central square. Connecting up the four cores were eight main link beams over the doorways. It was a logical layout, avoiding torsion, and as the stiffer cores resisted earthquake forces and became inelastic, there was damage, but it was controlled, and the building had a dual system and could also reduce the developed seismic forces as its period lengthened.

Right across the street was the Banco Central. It had a core at one end of a long rectangle, so it introduced torsion. That building was badly damaged.

The lessons learned from the 1977 Caucete earthquake have not been fully understood in Argentina. It is about 950 kilometers, 550 to 600 miles, from San Juan to Buenos Aires. Buenos Aires is such a large city that they have many, many buildings about ten to twelve stories tall. They are often arranged with parking at the ground story, with that level supported by columns, and apartments above with structural walls. When the earthquake in 1977 happened, there were people in these buildings who evacuated and ran away. There was furniture in the top stories that was moving around. About 50 kilometers from there, just a little closer to the earthquake, there was a large, new water tank for a factory complex. It was a steel structure of the mushroom configuration. The tank collapsed and burst open, and the rush of water pushed the cars against a row of eucalyptus trees and smashed them—in other words, like a tsunami. But still today, the university at Buenos Aires does not teach earthquake engineering.

Reitherman: Is this long-distance earthquake a potential problem in Buenos Aires as it was in Mexico City in 1985?

Bertero: Exactly. Mexico City was 350 to 400 kilometers away from the coast where the earthquake was released, yet buildings similar to the ones I am describing in Buenos Aires collapsed, if they were in the soft soil areas, on the lakebed. The national university, UNAM, Universidad Nacional Autónoma de México, is located on firmer soil, and the severity of the ground motion there was much less. You cannot just draw a line around where the earthquakes are released and say you do not have to worry too much about that hazard farther away.

There was an earthquake of over magnitude 8 in Bolivia in 1994, for example, that was very deep, and that was felt throughout Latin America and even as far away as the United States. For some structures and sites, these very distant earthquakes can cause problems, but people are not paying attention to these risks. In Argentina and in other countries, it is difficult for people who are far from the zones of frequent seismic activity to appreciate these risks. It is the same thing here in the USA. It is difficult to get engineers outside the western United States to take the seismic risk seriously.

Mexico City had damage from an earthquake in 1957 that was severe enough to cause some collapses of multistory buildings, but it was not given enough attention. That was a warning, almost thirty years before the 1985 earthquake.

I was in Mexico City in 2005 for the twentieth anniversary meetings about the earthquake, and there are still good people doing research there and trying to get the information to the practitioners. There is an architecture professor at *Universidad Nacional Autónoma de México*, Jesus Aguirre Cardenas, still working on the problems, for example.

Economic Pressure to Select Structural Systems

Bertero: For economic reasons, people want to build tall buildings with the structure concentrated in the middle, in the core walls around the necessary service areas of the elevators, duct shafts, stairs, and so on. And they want to leave the perimeter open for the views and light. But for seismic resistance, you would prefer the opposite—to put your structural elements around the outside, not pack them in the center. The ideal is to put your shear walls at the corners, or to have a seismically resistant frame around the perimeter, or both. That way, the structure resists torsion and overturning moments most efficiently. Historically, because you need the core and its walls anyway, you end up with the dual system of moment-resisting frame around the perimeter, and a core made up of ductile structural walls in the center, a redundant and reliable basic system when properly designed and constructed.

Now, however, there are tall buildings being built, even in a seismic area like San Francisco, that rely only on the core walls for lateral seismic resistance. Around the perimeter is a gravity frame, a frame designed to resist only vertical gravity forces. Now we have diaphragms that have to transfer all their lateral forces to the core, but again for economic reasons, these floors are very thin post-tensioned reinforced concrete slabs. The thin slabs are supported only on the perimeter columns and the walls of the central core, and those connections are difficult to make, especially because of the incompatibility of the vertical deformations between the perimeter columns and the walls of the core during their lateral deformations. As the building deflects laterally back and forth, we end up with a complex pattern of tension, compression, and shear stress in the diaphragm, and we will have cracks in concrete from the tensile stresses as well as some spalling and crushing from the compressive stresses.

Because the lateral force resistance of the vertical elements is only in the core walls, the use of an R Factor such as 4.5 to 5.5, as is presently allowed in the building code, implies significant

inelastic behavior there. The R Factor that makes the design economical is also a sign that the concrete will crack, crush, and spall, and that steel reinforcing bars will yield and buckle in the critical plasticized regions, also called the plastic hinges, through the complete cross-section of the core walls. This can create what is called sliding shear, a problem that is not always adequately tackled in engineering practice. Remember that if a steel bar is being stretched inelastically in an earthquake, only the small amount of elastic deformation is recovered as the loads and deformations keep cycling. The bar becomes longer than it was. In effect, it holds the crack open practically throughout the compressed cross-section of the wall. Thus, as the forces pull the wall sideways, it begins to lose its usual reinforced concrete resistance to shear, and instead the shear is resisted by only the dowel action of the vertical reinforcing bars that bridge across the crack.

Another problem is the inelastic behavior at the critical regions in the walls that could lead to a significant movement of the center of rigidity of the core and, consequently, significant increase in the inelastic torsion of the entire building. Remember that the perimeter, where resistance to torsion would be most effective, has not been made part of the lateral force-resisting system.

A basic problem in earthquake-resistant reinforced concrete design is to be able to actually construct what has been designed. There is a tendency to use as little concrete as possible, to have thinner slabs for example. And yet, the calculations may call for a large amount of reinforcing, especially at critical connections. This can be a great problem for the contractor to accurately provide the reinforced concrete material where it is called for—to fit the bars in but provide minimum spacing and cover, and to place the concrete. A line on a drawing that represents a reinforcing bar is easy for the computer to draw, but it may be difficult to construct. And a design is only as good as the way it is constructed.

In the lab at MIT, when I was a graduate student conducting experiments on beams and walls, I would order a precise mix of concrete to get 4,000 psi concrete. The concrete company would deliver 5,000-pound concrete. I would explain I needed 4,000-pound concrete, because that was what the specimens were designed for, what the intent of the test was, and they would say, "What are you complaining about? We are giving you more strength." That happens in actual building construction, and getting materials that are stronger than you assumed can negatively affect the behavior. It can change the location of plastic regions. It can reduce ductility that the engineer or the building code assumed.

The provision of construction materials that are overly strong as compared to the design assumptions is a current problem, as well as an old one. In Argentina from 1950 to 1953, when I was designing structures with our firm Weder-Bertero, there was a shortage of steel. A company began to produce twisted reinforcing steel bars. A bar was twisted enough to strain-harden it. The company was quite proud of their product—it had a higher yield point. But they had used up some of the original physical ductility that I wanted to rely on in my structures.

Soil-Structure/Adjacent Structure Interaction

Bertero: There is another adjacency problem in urban areas, beside the problem of pounding, which is lateral impact of adjacent buildings, or one building collapsing onto another. We do a soil-structure interaction analysis of one building to be built, as if it is in the middle of a huge, empty field. But right next door on all sides there are other large buildings experiencing soil-structure interaction. An engineer analyzing one of those buildings would also look at it by itself. That would neglect the probable effects that the response of the nearby building can have on the building being designed. The motion of the adjacent building will interact with the soil and thereby affect the ground motion input to the building under design. And the seismologist and/or geotechnical engineer today will give you the same ground motion as input to each of the buildings in an urban area without considering the soil-structure interaction from each of the other buildings that will modify that input motion.

Reitherman: Something like the way one boat that is set into a rocking motion affects the water around it and can rock the boat that is moored next to it?

Bertero: Yes, the ground motion is different because of adjacency. The rocking of a big building's foundation imparts its own motions into the soil, in addition to what the earthquake delivered in the first place.

Reitherman: How would you take urban-scale soil-structure interaction into account, especially if you don't know what might be built next door in the future? In wind design, there are wind exposure categories, where surface roughness category B applies to a site where the wind is reduced by numerous adjacent buildings and trees, C to a lesser extent, and D where the surroundings are flat. Would you zone urban areas in some fashion like that to take into account the seismic soil-structure effect of surrounding development?

Bertero: This is a very complex problem whose solution would require the cooperation and collaboration of experts in several areas of seismic engineering, such as a seismologist, geotechnical engineer, structural engineer, urban planner. Something has to be done to change present practice, which is based on just specifying the design earthquake ground motions based on records obtained in the free field. That neglects the effects of the soil-structure interaction that can occur in blocks of a city that are crowded with buildings that have completely different mechanical characteristics and foundations at different depths. For example, let us consider the case of a two- or three-story building with its foundation at practically ground level, and that it is surrounded by tall buildings with multi-level parking basements. The question is, what should be the earthquake ground motion used in the design of these buildings? Should it be based on free-field recordings, as is usually assumed? I do not think so. It is necessary to consider the effects of the soil-structure interaction, not only of each of the individual buildings, but the effects that such interaction can produce among all the surrounding buildings. It is my understanding that Professor Jonathan Stewart at UCLA and Professor Jon

Bray at U.C. Berkeley have started to investigate the effect of such interaction.

Reitherman: Could you give an example of a bad combination of adjacency soil-structure interaction, as distinct from a case where building response might be lessened by the interaction?

Bertero: For one building on its own, we usually consider soil-structure interaction to be beneficial. Radiation damping, like other damping, reduces the response of the structure. But let us now consider the case mentioned before of tall buildings with multi-level basements surrounding the two- or three-story building. Where does the energy radiate from the perimeter walls and foundation? It radiates away from the foundation and walls of that building, and toward the foundation and underlying soil of its neighbors. The tall buildings, with long periods, can hurt the short buildings. We should have learned more about this problem and applied lessons learned from the 1985 Mexico City earthquake.

Design for Aftershocks

Bertero: Let me tell you another lesson about ground motion that we should have learned from the large 1960 Chile and 1985 Mexico earthquakes. I think Allin Cornell is starting to look at this problem, but the codes, and most of the experts studying strong motion seismology, and the practicing engineers, do not consider the fact that the building that undergoes significant earthquake shaking is likely to undergo significant aftershock shaking. The 1985 Mexico City earthquake was not actually a single event. Even before you consider aftershocks, what is called the main shock may consist of multiple slips, seconds apart, with their ground motion overlapping. The 1960 earthquake in Chile had massive aftershocks, but even at present the engineering seismologist and/or the geotechnical consultant gives the structural engineer a design earthquake that represents only the main shock portion of what can be a multi-shock event.

Reitherman: What design approach would you suggest? For a long time, there has been the concept of a two-level design, considering a smaller, more frequent earthquake and designing the building to resist it essentially elastically; and then a separate analysis of the large earthquake, with the structure mobilizing its ductility for that case. Would you propose a complementary two-step analysis for the main shock, on the one hand, and for the aftershock or aftershocks, on the other, whereby the undamaged structure is analyzed for the large earthquake, and then the structure in its now-damaged state after that response is modeled for the analysis of one or more aftershock earthquakes? You would not only change the design earthquakes, you would change the structural model to take into account the residual damage and deformation it would have experienced.

Bertero: Yes, it is necessary to consider the changes in the mechanical characteristics of the model that is used due to the damage that has already been suffered. This concern is one reason for the current interest in structural systems that can re-center and recover after one earthquake, like prestressed columns that can rock and return to their original position. It is the problem of designing not only the structure, but the entire building system,

with sufficient hyperelastic resilience to avoid residual deformations.

Recently, there was a test of a prestressed column representing part of a viaduct here at the Richmond Field Station shake table. After its first earthquake simulation, with shaking corresponding to a code's maximum credible earthquake, the damage looked to be acceptable. However, when it was then subjected to a lower level earthquake, it went to the collapse point. You recall that a special crane had to be driven into the lab to remove the specimen. The good performance in the first earthquake was not proof that it could withstand the second one, even though the second one was smaller.

R Factors

Bertero: The building code specifies a constant reduction factor, or R, independent of the period of the structure. This does not make sense. Furthermore, it has usually been considered that it represents the benefits in reducing the required elastic strength due to the ductility, because the ductility ratio, μ, is often taken to be the same as physical ductility, which is not true.

Reitherman: What is the distinction between physical ductility and a ductility ratio?

Bertero: Consider the simple case of two identical buildings that will have practically the same reactive mass and which will be built near each other, that is, we assume the same soil conditions and design spectra. Let's assume the spectrum has the same maximum demand in the period range from 0.25 to 0.50 seconds. Assume that it was decided to use the same structural system, but using different lateral stiffness. The stiffer structure has a period of vibration, T_1, of about 0.25 seconds. For the less stiff building, T_2, the period equals 0.50 seconds, which indicates it is four times more flexible than the other—because if we hold the mass constant, the period varies inversely proportionally to the square root of the stiffness. It follows that the lateral deformation at yield for the more flexible structure is four times that of the stiffer one. Yet the ductility ratio, μ, of the amount of deformation at the ultimate limit deformation compared to the deformation at yield can be four in both cases.

The development of physical ductility leads to damage of the structure. The larger the amount of this physical ductility, the larger the amount of damage. To control damage, we must control the amount of physical ductility. Yet here we have a case where the ductility ratio, μ, is the same for both structures, while the physical ductility, and thus the damage, is four times larger for the longer period structure, which can fool the designer. The R Factor for both structures is the same, the μ or ductility ratio is the same, but the damage we would expect is quite different.

When people say "ductility" you have to question them as to what they exactly mean. When people refer to the R Factor, you have to question them as to what they understand by that term. Yes, the building code has to be simple. But the code also has to be reliable.

Back when ATC 3-06 was developed, which was essentially done by 1975, I had reservations about the R Factor concept that was developed in that project. When you see the word "equivalent" used in seismic design, you must take a close look at what it really means. It is true that a structure can develop its ductility and damping to resist earthquake forces that seem overwhelming in an elastic context. It is desirable to have a simple analysis of elastic behavior that is equivalent to the actual inelastic behavior. If you increase the damping in the elastic analysis, you reduce the calculated forces. But you have to understand the physical behavior, not just put different factors together to come up with calculation results that you want. You can't just say the damping is "equivalent" just because it adjusts the calculated forces to reduce them to approximately the right range as compared to the structure's capacity.

If you look at the equations of energy, you have fluid damping, but you also have damping dependent on velocity—two different things. A push-over analysis does not include velocity, but it includes terms for damping. A push-over analysis is a static type of analysis in which the structure is assumed to experience a particular drift, as if it were "pushed over," and the behavior of the components of the structure are compared with that level of motion. In a pushover analysis, the engineer wants to select damping values that will make the overall forces seem appropriate, but you have to understand when the calculation method departs from the real behavior of the structure.

Engineering Versus Economics

Bertero: There is often a conflict between achieving a high level of toughness and hyperelastic resilience on the one hand, and keeping the cost low on the other.

Of course, each building is different and you have to look at each one. In the 1970s I started

105

looking at the buildings in general in the San Francisco Bay Area, and what I found was a pattern where it was not the short buildings I was worried about for a future earthquake, and not the very tall ones. It was the ones of medium height—four, five, six stories, up to maybe about eight stories. Many were developed as products to sell on the market, like any product. Condominium projects are often like that. The buildings are often designed differently than the way engineers design a corporate headquarters or government building that the owner will own for decades or a century. Economics and engineering can be in conflict. Initial cost can be made lower, but that means that when the earthquake occurs, the cost suddenly jumps up, to repair or replace the building. The developers do not own the building for the long-term, and they will not be the ones to pay for earthquake damage in the future, so it is in their interest to minimize their up-front costs.

Now, we have a trend in many cities toward tall condominium buildings. It is the same conflict, between business interest and the engineering to adequately protect the building from earthquakes that may occur sometime in the future.

We also have a problem with the economic pressure on the consulting engineers. ATC 3-06 was produced only because of very large amounts of volunteer engineers' time. You could not do that today. Consultants are under more competitive pressure to do paying work.

Performance-Based Design

Reitherman: Theoretically, performance-based design could be a technique to design a structure or facility to meet any performance level, from just barely preventing collapse to prevention of even minor damage. In practice however, both the researchers and the practitioners involved with performance-based seismic design are aiming at producing—and increasingly they are promising—seismic designs that will predictably attain high goals, such as keeping the building functional and the repairs nominal. Aside from the extra cost to meet a higher performance level, what do you think of the analytical complexities in predicting those higher levels?

Bertero: I'm not sure many designers can reliably carry out performance-based design. In 1995, I thought it would take 15 years to reach that goal. In 2002, I reviewed the work on this topic that the Pacific Earthquake Engineering Research Center (PEER) has done, and then I still thought it will take 15 years. You cannot change the profession of structural engineering overnight. Elastic-basis design with appropriate detailing and so on for ductility may deliver reliable structural safety, but it is not the same thing as reliably predicting exactly how the entire building system will perform.

The design engineer may be asked to select four or five levels of earthquake ground motions and the corresponding levels of deformations, for example. It gets complicated, and yet is still a shortcut. The level of education of the profession needs to be elevated considerably.

Probability

Reitherman: You have commented on the need for engineering students to have first-hand familiarity in the laboratory with the testing of materials. Does that learning experience extend toward helping them understand how much uncertainty there is in the capacity of the structures they design?

Bertero: There is uncertainty in the design spectrum. The earthquake that shakes your building may be significantly different. The design spectrum is an estimate, not an exact prediction. You have uncertainty in the dynamic behavior of the different components of the overall structure and of its connections. There are also a whole set of uncertainties in the behavior of the nonstructural components of the building system, and contents in the building can also be very important.

When I started to teach the undergraduate course in the design of reinforced concrete, I would have the students test three small beams. They had to design the beams, make the reinforcement and formwork, design the mix of the concrete ingredients, make the concrete by combining those materials, pour and cure the concrete. The students had already had a course on structural engineering materials, but I felt that to learn reinforced concrete design they had to appreciate how its material properties could change. A small change in the formwork precision, for example, could affect the strength. The students could understand what makes up the uncertainty when they were taught this way.

They took samples of the materials and tested them first. Then the students, having made their beams, tested them for severe concentrated loads and predicted the response of them. The response was in terms of the forces in the beam versus slowly applied static loads. They had to consider the different levels of deformation or damage that the beams would undergo: first flexural crack, first diagonal shear crack, first yielding of the steel reinforcement, first spalling of concrete, deformation at the maximum resistant force, and maximum deformation at incipient collapse. Their final reports had to show what they had learned about comparing analytical predictions and observed behavior.

I am really concerned about the way the exteriors of tall buildings are being built at present. I would not want to be on the sidewalk in downtown San Francisco in an earthquake. In the structure, we recognize the need for redundancy—what happens if this column or that column fails? Some of the cladding panels are held in place in a very nonredundant way with a small number of connections. Enough testing has not been done on the behavior of the panels and of the glass at the corners. You have to imagine drift at the corner of the building, where the drift along one side of the building is incompatible with the drift along the other.

Energy-Based Seismic Analysis

Reitherman: Engineers almost always calculate seismic loads in units of acceleration times mass. But back in 1956, George Housner clearly stated the more fundamental energy basis of the seismic response of a structure: "The effect of the ground motion is to feed energy into the structure. Some of the energy is dissipated through the damping and the remainder is stored in the structure in the form of kinetic energy of motion of the mass and in the form of strain energy of deformation of the structural members…. If the structure is designed so that permanent deformations can occur without failure of a member, then at any instant the sum of the kinetic energy, plus strain energy, plus energy dissipated through

107

normal damping, plus energy dissipated through permanent deformation will be equal to the total energy input."[52]

Along with Housner, your name is associated with early research on using energy as the essential parameter in seismic design. How did you get involved in that concept?

Bertero: Professor Housner was ahead of his time. You can directly convert the equation of motion into the equation of energy. It gives you the complete picture. Sometimes practicing engineers incorrectly say that plastic behavior is equivalent to damping. In the push-over method, it is important, conceptually, to know that you have a certain amount of energy dissipation and how it happens.

If you're going to move to innovative seismic design, there are two things you can do. One is to try to understand the plastic behavior—the actual ductility, not the ductility ratio, as I explained earlier. The other is to understand damping. You either reduce response by ductile behavior of the material, or by damping, such as by adding damping devices. These are two different concepts, both based on an understanding of the actual behavior of the structure. And that understanding is less today, because the engineering schools have less laboratory instruction. A deformation applied

slowly is different than that same deformation applied rapidly, which the energy concept clearly indicates. Applying R Factors with elastic-level forces is not the reality.

Reitherman: When did you start investigating energy as a central concept?

Bertero: My second doctoral student was James Anderson, now a professor at the University of Southern California. He received his PhD in 1969. We worked on the energy approach back then, and Jim has continued that line of research.

Reitherman: Do you think an energy-based seismic design method will evolve that will have widespread application by practicing engineers?

Bertero: Yes, with time. It will have to be economical of the engineer's time, because of competitive pressures. The consulting firm needs to produce designs quickly. And to change engineers' methods always causes them to do more work, as compared to doing it the same way they have typically used in the office.

Unfortunately, if it's not in the building code, most practicing engineers will not do it. Engineers need to be taught new methods, and then gradually they will be adopted in practice.

---

52. Housner, George, "Limit Design of Structures to Resist Earthquakes," *Proceedings of the World Conference on Earthquake Engineering*, June 1956. Earthquake Engineering Research Institute, Oakland, California, 1956, p. 5-4.

Chapter 11

Family and Friends

In Argentina, we were used to the parents having much more control over the children. I don't know if that is good or bad, but it is my observation that it was very different living in Berkeley.

Reitherman: You've mentioned your family at various places in the preceding narrative, but please name your children here.

Bertero: Nydia and I have six children. In order from oldest to youngest, we start with María Teresa, nicknamed Teresita, who today works for the system-wide academic senate of the University of California, at the state headquarters of all the campuses in Oakland. Edward, nicknamed Eduardo, received a degree in business administration and is in the re-insurance industry. María Teresa and Edward were the two children who were already born when we moved from Argentina to Massachusetts, when I went to graduate school at MIT.

Then comes Robert, who has a master's degree in public health and is a paramedic expert. Mary Rita studied business administration. Adolf has a master's in public education. And then Richard, the youngest, took some courses in engineering, but none of the children ended up becoming an engineer. They all went to college in the United States.

Now we have ten grandchildren, and one great-grandchild. My wife is very devoted to them.

Reitherman: What kind of family vacations did you take?

Living in Italy for a Year, Vacations

Bertero: In 1964 and 1965 I had my sabbatical and spent time with my family in Europe, based in Italy. I went to work with Professor F. Levi, who was the Director of the Institute of Construction Sciences at the University of Venezia, *Università Ca' Foscari Venezia*. Professor Levi was doing work on creep in concrete, the rheology of the material, and I was interested in that. In 1964, in the ASCE and ACI International Symposium on Flexural Mechanics of Reinforced Concrete, which was held in Miami, Florida, I met Professor Giorgio Macchi who was working with Professor Levi on the redistribution of moments in reinforced concrete structures due to its inelastic behavior. Professor Macchi helped me to be accepted by Professor Levi to work for one year at their institute in Venice.

One assistant to Professor Levi gave me a very nice tour of several buildings in Venice where apartments could be rented. The buildings were beautiful, they were basically former palaces, with paintings on the walls and ceilings. But when my wife looked at the bathrooms

and kitchens, you could see how hard it would be for me and my wife and six children to live there. We ended up renting a place in a new building in Vicenza, about a 40-minute train ride away, and I would take the train everyday to Venezia. I think the assistant to Professor Levi thought we did not appreciate art and architecture, but I explained that in the United States, the wife has no servants to do all the work. My wife was going to take care of the household and children, and she needed a good place to do that.

Reitherman: The train station in Venice ends at the water's edge. When you got there did you set off on foot, crossing little bridges to make your way to the university every day, or did you go via canal on the vaporetto?

Bertero: If I was in a hurry, I would get in the boat. It cost 100 lire. One day the boatman said it was 500 lire. I responded to him in Italian and he realized I knew he was trying to overcharge me. That aspect of living in Italy was not pleasant. My wife and I would go in a store to buy the children clothes and they expected you to negotiate the price. You never knew what the real price was.

I traveled from Italy to visit faculty members at the University of Cambridge, Professors J. F. Baker, M. R. Horne, and Jacque Heyman, the steel experts. We went to have lunch in their faculty club, and I could not believe it! I had to borrow a gown.

Reitherman: You mean you needed more than a coat and tie to gain entrance to have lunch?

Bertero: Yes, it was very conservative. And when you were at the table, you had to wait to sit down at the right time. In 1967 at Berkeley, things were quite different!

We visited Torino, or Turin, where there were relatives of my father, as well as the small village where my grandfather was born. This in the Piamonte (or Piedmont) area of Italy. In small villages in Italy, nothing had changed. On Sunday, the women would go to church and the men would go to the bars to play cards and bocce. It was like the customs in Argentina, where the Italian immigrants brought the old ways with them. We also visited relatives of the father of my mother. They were living in the house where my grandfather was born. This house was located in a village located near the border of Italy with Yugoslavia, that is, near Udine and Trieste. Even at that time, people would go into the mountains and come back carrying on their backs tremendous amounts of firewood to store up for the winter.

During that year, I traveled with my family all over Italy—south, north, east, west. I think it was an education for my children. They learned a great deal, and it changed the attitudes of some of them toward school. They realized it was valuable to learn languages. The older two children, Teresita and Eduardo, spoke Italian fairly well.

During 1971-1972, when I was appointed as the Chief Technical Consultant of the UNESCO mission at the IISEE in Japan, part of my family came with me, so that was another interesting experience for them. The three oldest children stayed in the U.S. because they were already in college then.

Here in the USA, we took vacations, usually a week long, to Yosemite or Lake Tahoe, and sometimes to Mexico and Canada.

Living in Berkeley

Reitherman: Except for the years abroad in Italy and in Japan, you have lived in the city of Berkeley since joining the faculty of U.C. Berkeley?

Bertero: Yes. In the 1960s in Berkeley, lots of drugs were used by young people in the public schools. We sent our children to the School of the Madeleine and to Catholic high schools in Berkeley to avoid that. In Argentina, we were used to the parents having much more control over the children. I don't know if that is good or bad, but it is my observation that it was very different living in Berkeley.

It is not possible for someone today who was not in Berkeley before the political changes of the 1960s to realize what Telegraph Avenue was like back then. It was a very nice street for families. My wife would take her friends who visited to have an espresso and visit the shops along the street. I don't know what is a good change or a bad change in general, but for the children, it was not a good change. We were strict on where they could go. They would say, "Dad, can I go to this dance or this party?" I would ask, "With whom will you go?" We did not let them go alone when they were younger if we didn't know their companions.

Chapter 12

A Conversation With Former Students

I am glad to see you all be leaders now. I have confidence for that reason in the generation you have taught, the ones who will carry on.

[On April 28, 2008, a small group of former students of Professor Bertero met with him at the U.C. Berkeley Earthquake Engineering Research Center in Richmond, California, in an informal roundtable discussion as part of this oral history project.]

James Anderson

Anderson: I had been at Berkeley in graduate school for about a year, and then I came into contact with Professor Bertero's classic course on dynamic loads.[53] When it

---

53. James Anderson was a PhD student of Professor Bertero from 1966 to1969. His thesis topic was *Seismic Behavior of Multistory Frames Designed by Different Philosophies*, published as U.C. Berkeley EERC Report 69-11. He is a professor in the Department of Civil and Environmental Engineering at the University of Southern California.

came time to look for a thesis advisor, I thought he would be a good prospect. I finally looked him up. At that time he was in a back room of the old wooden architecture department building on the north side of campus. If you went down a hall or two, through a big classroom, into a glassed-in porch, you could find Professor Bertero behind a tall stack of books.

Later, he moved into Davis Hall, and I remember spending a lot of time outside that office, waiting for my turn to get in. Every week, you had to be ready for your meeting with him, and that motivated you to keep working. I'm not quite that hard on my students.

In my thesis work, we wanted to look at different techniques used to design buildings to resist earthquakes. At that time, the standard method was the building code method on an allowable stress basis. Then there was a method using some plastic analysis, and we had a method based on minimum-weight design, which led into energy-based design. One of the best methods was what we called the strong column-weak girder design. I took an old computer program that Professor Ray Clough and Professor Ed Wilson had developed—originally, I think, to look at some buildings that were damaged in the 1964 Alaska earthquake. I modified it into a more general program for building analysis. We did quite a bit of nonlinear analysis.

Reitherman: You started right then in your doctoral work on the energy-based approach, which you have continued through to today?

Anderson: Yes, though my work then on the energy approach was rather minimal. We were using energy balance equations not to do the design, but to check the analysis and to make sure our analysis was not going off into left field somewhere. We tried to keep track of the energy quantities and keep them in balance. That's a little bit different than what we are doing today.

I don't think energy approaches are going to permeate into practice very soon, but it is picking up speed in terms of the number of people doing research in this area, especially since roughly 2000. It has some distinct advantages, such as being able to account in some sense for duration, for example.

Reitherman: When you went on to become a professor at the University of Southern California, did you use "Bertero-esque" teaching methods?

Anderson: I don't know if anyone could really do that—it would be hard to try to copy him. But I did catch on to one of his techniques. He used to give out his own handwritten notes in his classes. He used to struggle with those mimeograph machines. I do the same thing in my classes, but now I have access to a nice photocopy machine that whips them out in no time. Another technique I have carried over—I always go everywhere with a red pen. [Laughter. Anderson pulls a red pen from his shirt pocket]. You see, Professor Bertero is legendary for his use of red ink, to mark up papers, or make red bullets on his overhead transparencies.

Reitherman: Jim, you overlapped with the time when Helmut Krawinkler was a graduate student—did you get to know each other then?

Anderson: Oh yes—we both spent a lot of time waiting in the hall together [laughter].

Helmut Krawinkler

Krawinkler: I came to Berkeley in 1966[54] and had no contact with Professor Bertero for about half a year. Then, I signed up for Professor Clough's dynamics class, but it turned out Professor Bertero was teaching it that year. I arrived promptly on time—which was a big mistake. If you arrive on time for a Bertero class, you are already late. By the time you get there on time, the whole blackboard—a big blackboard—is filled with tiny white chalk writing and diagrams, from one end to the other, which you quickly start to write down. And then, on time, he starts talking and you have to take notes on that too, and you never catch up.

That was my first mistake—to arrive on time for a Professor Bertero class. My second mistake—not really a mistake, but many of my colleagues at the time considered it a big mistake—was to take on a research assistantship with Professor Bertero. They said, "He will work you to death," but fortunately he only succeeded halfway. And then, secondly, my friends told me, "He will raise his voice with you and scare you." The right approach is, first to let him shout if it's not important, and if it's

important, to shout back. It turned out to be a good relationship.

As Jim Anderson said earlier, we spent a lot of time in the hall waiting our turn, and one reason was that Professor Bertero was consistently late. He operated on Latin time, and Latin American time has little to do with real time [laughter].

My colleagues also considered it a mistake, but it was my good fortune, to be a research assistant to two professors: Professor Bertero was my primary one, and Professor Popov was my second advisor.

Those two professors had an interesting relationship. May I tell the story, Professor Bertero? Okay, it happened a few times, that these two people, each of whom had a very strong but very different personality, didn't get along. I was a PhD student, but I was also a "marriage counselor." Professor Popov would tell me, "Let Professor Bertero know that we have to do this research this way," and then I would go to Professor Bertero, who would say, "Tell Professor Popov we need to do it this way." I got my exercise going back and forth. They were a great team, but sometimes their personalities were just a little too different. They were the best of friends, but like your own wife, there may be times when you argue.

Bertero: Professor Popov did analytical-experimental work. It was all together. When Helmut did his work with us, first he had to review everything that was known on that topic. Then what? Analysis, analysis, analysis. First the analysis, then the instrumentation, and only then you do your test. Helmut's work created duplication—triplication—of ways of measur-

---

54. Helmut Krawinkler was a research assistant as a graduate student from 1967 to 1971 for Professor Bertero and Professor Egor Popov, finishing his dissertation in 1971, which was published as EERC report 71/07, *Inelastic Behavior of Steel Beam-to-Column Subassemblages*, co-authored with Vitelmo V. Bertero and Egor P. Popov. Then he did a year of post-doctoral work, again with Bertero and Popov. Helmut Krawinkler is a professor emeritus at Stanford University in the Department of Civil and Environmental Engineering.

ing things. He had to do different measurements of the same thing, to check the results. And then at the end, you do analysis again.

Krawinkler: That was 1968, 1969, 1970. Electrons existed but the electronic age didn't. The data was recorded point by point. You had pieces of paper on which you plotted points to get hysteresis loops. It didn't come out of the computer. You had to digitize the individual data points. Instrumentation was a challenge. Data acquisition was a challenge. And testing was a challenge, because we didn't have MTS-controlled actuators. We had three manual pumping systems. In the test setup, we had to apply synchronized axial load to the column, synchronized simulated gravity loading, which of course had to be kept constant as things moved. There wasn't any MTS device to do this for you. There were two or three of our good graduate students manning each of the manual hydraulic pumping stations. You looked at an X-Y recorder and had to holler "up" or "down" to them.

When a test was over, Professor Bertero was the first to go to the specimen, looking for cracks in the steel, even with a magnifying glass, looking for distortion in the joint panel zone. At that time, we graduate students were busy pushing buttons to record data, on 200 channels. We had more than 60 strain gauges in the joint panel zone.

Bertero: Sometimes there were seven or eight people working there, with Helmut commanding the whole thing.

Krawinkler: We used a theodolite, a surveying instrument, and took still photographs using glass plates, because film distorts too

much with temperature. We had about fifty to one hundred glass plates, about four inches by six inches, and I spent many days and nights going to Menlo Park where USGS had a comparator, a nice little instrument where you put in the glass plate with the photography showing the grid marked on the specimen, and then one point by one point I could digitize the points on the comparator. Then eventually, we re-drew the whole thing for the publication.

Anderson: You're probably one of the last projects that used a surveyor on your laboratory testing team.

Krawinkler: My graduate work was the cornerstone of my career for many years. The emphasis was not so much on the columns, or the beams, but the material between, the joint panel zone. It was a controversy in the design profession. It was standard practice that you had to design it so that it was stronger than the connected beam and column elements, which meant you needed a big, heavy doubler plate in the joint. This was one of the most expensive aspects of constructing a steel earthquake-resistant moment frame. It was a very touchy issue. We went from very strong panel zones, to weaker panel zones.

The whole idea was to distribute inelastic deformations among the elements so that they don't become too large on any one element. Alternatively, you can concentrate all your deformation in one place, the joint panel zone, in which case it becomes a parallelogram that distorts a heck of a lot, and you get kinks at the corners, which in turn causes problems at the welded connections. It's a balancing act.

Reitherman: On February 9, 1971, the San Fernando earthquake happened early in the morning. That would have been during your doctoral work. Do you remember anything about the news of the earthquake then?

Krawinkler: Yeah, there was excitement. I went down there, but I had very little time because it was at a critical point in my work on my dissertation. Earthquake engineering up to then was not really that popular a research topic. Earthquakes happened in distant places, but not here in the United States. The San Fernando earthquake of 1971 brought the message home.

Before then, there wasn't enough money from the National Science Foundation (NSF), the American Institute of Steel Construction (AISC), or the American Iron and Steel Institute (AISI) to do experimental research on steel structures and earthquakes, which is expensive. After San Fernando, there were more funds to do that sort of research. That disaster memory lasted for about ten years, and then faded out. Funding has been declining in general, in real dollars, with occasional upward kicks.

I wasn't even planning on staying in the earthquake field. I came here from Austria as a Fulbright student. I was planning on returning to Austria, but there wasn't much future there for someone with a PhD in earthquake engineering. I also need to give my professors at San Jose State University, where I got my masters, the credit for re-igniting my interest in structural engineering. I had lost interest in it in Austria because my instructors tended to be behind the times.

So, with my doctoral work in earthquake engineering, I was lucky to get a Stanford position, and I have been there for thirty-five years. I am retiring this year.

Reitherman: When you joined the Stanford faculty, who were the other earthquake engineering people there on the faculty?

Krawinkler: Primarily it was Haresh Shah. Jim Gere was doing some things in dynamics, but not specializing in earthquake topics. It was primarily Haresh. The position that I came into was vacated by Jack Benjamin when he left Stanford to do consulting engineering. I would say I "came into" his position. I would never say I "replaced" Jack Benjamin.

Stephen Mahin

Reitherman: Let me ask Steve Mahin[55] a question. Are you the former student here who spent the most student years at Berkeley? For example, Helmut had his undergraduate education from the Technical University of Vienna and a master's from San Jose State. Jim Anderson did his undergraduate and master's work at the University of Michigan.

55. Stephen A. Mahin did his MSc work 1968-1970 at Berkeley and his PhD with Professor Bertero as his advisor 1970—1975, then did post-doctoral research 1975-1977. His thesis topic was published as EERC report 75/05, *An Evaluation of Some Methods for Predicting Seismic Behavior of Reinforced Concrete Buildings*, co-authored with Vitelmo V. Bertero. He is a professor in the Department of Civil and Environmental Engineering at the University of California at Berkeley.

Mahin: I started here in 1964 as an undergraduate, and I've been here ever since.

Reitherman: When did you first meet Professor Bertero?

Mahin: I took a strength of materials class from him. I was influenced by Joe Penzien, Egor Popov, and Boris Bresler, but Vit was the primary influence on my career. Helmut was my teaching assistant in that first class I took from Vit, though I don't remember much about that.

Krawinkler: Well, I remember you, Steve—long curly hair [laughter]. But I didn't know his last name was "Mahin" [May'—in] because the only person who mentioned his name to me was Professor Bertero, who called him Mah-heen [laughter].

Mahin: In all my classes I used to use the little lab books, like you would use in chemistry. I would fill up half of one, or maybe two-thirds of one. But in Vit Bertero's class, it was the only time I filled up three of them. He had office hours in the same room Jim Anderson mentioned, in the original campus architecture building near Davis Hall. At that time, Wurster Hall had just been built to be the new architecture building, and the old wooden one on the north side we're talking about was a naval architecture building.

I had worked for four or five years as an architectural designer, and for Hewlett-Packard as a product designer, and didn't particularly like either job. That was when I met Henry Degenkolb and worked for his office while going through school for a while.

I started working for Degenkolb in the summer between my junior and senior years, then worked halftime for them for two years. I moved over to a research project on the 1967 Caracas earthquake that SEAOC was doing, which involved Paul Fratessa, Harry Seed, and others, to see if this new dynamic analysis would ever catch hold and prove its worth. It was forensic analysis, to be able to predict why buildings had collapsed.

Whenever Henry wanted something done, he would come out of his office, look at me and another junior guy, and usually pick the other guy who could turn out drawings quicker. But eventually he came out of his office with a cigar in one hand, and maybe at the end of the day a bottle in the other [laughter], and picked me. In a way I became Henry's guinea pig to try things out, and I worked for Loring Wyllie and others there on miscellaneous things.

I was looking for something to put a lot of energy into, and what I learned from Vitelmo was that if you spend time at something, it should be something important, something you can be passionate about and throw yourself at. I appreciated his intensity at whatever he did. In my PhD work, I learned the habit of critical thinking. With my students now, I think to myself: just slow down, what is this student trying to do? What are the assumptions? Why are you doing this? What is the next step?

Like Vit, I like to study many different things, at the system level, on down to the detailed level of how the materials behave.

The other thing I learned from you, Vitelmo, was pushing things one step further. I remember I was working on my thesis, doing all the drafting in ink, getting handwritten work typed up. The day before my dissertation was due you said it was acceptable—except that it needed one more section. So I wrote an appendix. And now, I have the same tendency to try to push a little further than the minimum, even when the deadline approaches.

Reitherman: Isn't there a tale about the one and only original copy of an important manuscript getting tossed out accidentally and ending up in the city of Berkeley dump?

Krawinkler: No that's my story, how I went to the dump to search for my document.

Mahin: My story too [laughter]! I'll tell you my story and you can compare your story about the dump, Helmut.

This was when you did things by hand, and I had written the thick handwritten report on the Olive View Hospital.[56] I was standing there on the fourth floor of Davis Hall, by the wastebasket that had one of those swivel tops, and as I was waiting, I set the document down on top of the trash can. I got distracted talking to someone, then the elevator came and I dashed for it, then later realized I had left the one original document behind.

When I got back to the fourth floor, the document wasn't there and the trash had already been collected, it had been taken to the Berkeley dump. So I went down to the dump by the Bay, and it turned out that the garbage guy had taken

it out and thought it was interesting, so when I asked about a thick green binder he had it.

Reitherman: Helmut, you were saying you had a story about the Berkeley dump also? I've previously heard Steve relate the incident about the dump and his lost manuscript. Something similar happened with you?

Krawinkler: It is remarkable how lives cross. In the case of Steve and me, our lives crossed at the Berkeley dump.

The data for an experiment in the 1970s was printed out on strips of paper, and that represented at least half a year of our lives. All the strips of digital data—there was no hard disk in those days, there was no copy or backup—went into a big book. I left the document on top of the garbage can in the corridor of Davis Hall. I was wearing a suit because we were going to have a project meeting. I went back to get the big folder of data, and it was not there. I went out to the dump down by the Bay in my suit. They told me where the university's garbage had been dumped. I found all kinds of interesting papers of professors that I should not reveal [laughter], but not my folder with all the data. It was foggy by the Bay, sea gulls swooping down at me, it smelled, it was spooky. A terrible experience. After three hours I gave up. In the evening, I was back in Davis Hall and told the janitor my sad story, and he said, "Don't be sad, it's not lost. I took it for my kids. It was too interesting to throw out." He brought out the big folder and handed it to me. So that saved my PhD, if not my life.

Mahin: The work on the 1971 San Fernando earthquake kept me fully involved in the

---

56. Stephen A. Mahin, Vitelmo V. Bertero, Anil K. Chopra, and Robert G. Collins, *Response of the Olive View Hospital Main Building During the San Fernando Earthquake.* EERC 76/22, University of California, Berkeley, 1976.

earthquake area, and that was rapidly followed by the 1972 Managua and 1976 Guatemala earthquakes. I remember climbing up and down the Banco de America building in Managua endlessly, only eighteen stories, but if you use the stairs to go up and down a few times each day—the elevators were out, of course—you get tired keeping up with Vitelmo. And I managed to get dysentery, eating the food from the street stands. Vitelmo didn't get sick, but I did, for two weeks.

In my qualifying exam, I had close to 200 pages of response to two questions by Vit. There was a question about braced frames, and another to devise a long-term program for seismic research on reinforced concrete structures. In response to the latter question I proposed hybrid, computer-controlled testing. We should have been smarter and just published it at that point and gotten more credit. Today, people take hybrid testing for granted, but back then it hadn't been thought of yet.

The earlier part of my student time at Berkeley was in the Vietnam War era. I recall helicopters dropping tear gas to quell demonstrators while I was in T. Y. Lin's prestressed concrete class. During my post-doc years, 1975-1977, I was doing research and consulting with Egor Popov on offshore platforms for Shell and Exxon. I joined the Berkeley faculty in 1977. When the Imperial County earthquake of 1979 occurred, I worked on one of the first fiber-based modeling projects to look at the Imperial County Services Building. In the 1970s, every few years there was an earthquake that kept the researchers busy.

I must admit, as a home-grown American, that I learned English from Vit [laughter]. I had the casual American viewpoint about the English language—if it sounds good, it must be all right. To this day, I can't write "The results show that…." Vit taught me to say "From an interpretation of the results, one can infer that…."

Eduardo Fierro

Reitherman: I think Eduardo Fierro[57] comes next, chronologically. What do you remember about first taking a course from or meeting Professor Bertero?

Fierro: When I met Professor Bertero, he was 53 years old, and that seemed old. Now I'm 55 and I think that's a young age. Time changes your perspective.

I went to Notre Dame for my master's, and when I came to Berkeley, it was the only university I applied to because I didn't have the money to pay the application fees for more than one. Fortunately, I was accepted, but I had no money.

I needed money to pay for my tuition. I asked around, and no one seemed to have any research assistant funds. Some of the students said that maybe Professor Bertero had some research money, but, they said they did not recommend that I work for him. They said he was so difficult, so tough. He's going to scream at you. He's going to make you work so hard.

---

57. Eduardo Fierro was a graduate student of Professor Bertero at U.C. Berkeley from 1978 until 1981. He then worked for Wiss, Janney, Elstner until recently becoming the "F" in BFP (Bertero Fierro Perry) Engineers.

But I was desperate and I went to Professor Bertero, and between him and Professor Popov they hired me to do some research assistant work.

I have heard the other stories today about his doctoral students having to wait outside his door for a long time. Well, I am a Latin man, so I arrived late, and Professor Bertero would raise his voice with me for that. Sometimes when I met with Professors Bertero and Popov, Professor Popov was very intense and had fire in his eyes while Professor Bertero was more quiet. Sometimes it was the reverse. They wanted so much of you in just one week, and then they wanted more the next week. The research topic was the seismic behavior of the re-inforced concrete beam-column-slab subassembly. They had me read, read, read, and there was almost no research that had been done that had included the slab in the testing. Rules of thumb were used, like the influence of the slab extending for eight times its thickness—which was not true, it has more influence. There was no solid evidence behind conclusions about the influence of the slab. If it was important to have a strong-column, weak-beam condition, you have to realistically include the slab. I have been to many earthquakes, and although I am told it happens, I have never seen the beam hinge as it is supposed to do, and this can have to do with the influence of the slab.

In the lab, one of the first things I did was break an aluminum transducer. The welds were weaker than they should have been, and I learned how to temper the metal in the oven slowly. Professor Bertero taught me to be suspicious of welds.

He taught me to look very carefully at the specimen before and after testing. No one looks more carefully than Professor Bertero.

That was the best experience I ever had in really learning about earthquake engineering. I had taken earthquake engineering in the best university in Peru, but I realized I didn't really know much.

Professor Bertero would ask a question, and then ask you if you already knew the basic answer before you ran any numbers. What do you expect the results to be?

Every week, a student would make a presentation before other students. Every six or seven weeks it would be your turn again. That was one of the best learning experiences I had. You gain confidence and skill in explaining yourself. If you can present in front of Professor Bertero, you can present in front of any audience.

He also gave me my test to fulfill my language requirement. We talked for about five minutes in Spanish, and that was that.

In 1999, EERI wanted to put a reconnaissance team together to study the earthquake in Colombia. I ended up the team leader. Everyone had to be fluent in Spanish for two reasons. First, it was helpful to gather information, but it was also thought that it would make us blend in more and be less likely to be kidnapped. One of the pieces of advice Professor Bertero gave me was this. Every night, make your team members give you a list of things learned that day, ten bullet points. Every team member, every day, ten bullet points. At the end of your trip, you have your report written. I was also the cook, and one night, when one guy didn't give me his ten bullet points, I told him I wouldn't give him his dinner. So he gave me his list.

Reitherman: You were also the team leader for the EERI team studying the 2007 Pisco, Peru earthquake. When this oral history volume is published, you will read a story or two told by Professor Bertero about the beverage of the same name.

Fierro: It's a beverage, yes, and it is also a serious international controversy!

Reitherman: What do you recall Professor Bertero teaching you in graduate school that is still relevant guidance for you?

Fierro: He told me that the best observations you can make of a shake table test is to go see what has happened in a real earthquake. How many shake table tests do you see in your life? Ten? Maybe twenty? If you go to an earthquake, you see thousands of tests, conducted on a shake table maybe 40 kilometers by 50 kilometers.

Lately, Professor Bertero and I have been teaching in the Caribbean, in the Dominican Republic. In Turkey, we taught the same class, in English. We also did a lawsuit job, and after that he said "No more litigation—ever." The lawyer kept trying to lead him into speculating on hypothetical questions, and Professor Bertero just kept saying, "I will explain again what happened." And now, Professor Bertero advises for the consulting firm that includes myself and Cynthia Perry.

Reitherman: How much time passed between when you left Berkeley and when you

started working with Professor Bertero on these non-university activities?

Fierro: I never lost contact with you, did I, Professor Bertero? Maybe it is a Latin thing. We Latin American students thought of him as a father. We would visit socially.

In Latin America, Professor Bertero is not only respected, he is revered. I was in a meeting in Chile, and we were discussing an issue about reinforced concrete. They didn't believe what I was saying. I picked up the phone and called Professor Bertero, put him on the speaker phone, and said, "We have a question here." Professor Bertero gave his answer. I thanked him and hung up. Then there was silence. The Chilean engineers were astonished. "You called him, just like that? And he talked to you?"

James Malley

Reitherman: Jim Malley,[58] I think you come next in the chronology. Talk about your Berkeley days.

Malley: It was the fall semester of 1979, my senior year. I had a young professor for reinforced concrete, Professor Mahin. We went through about the first two weeks of class. Then we showed up for a lecture and Professor Mahin wasn't there. Another professor was there, already filling the blackboards with

---

58. James Malley received his master's degree in civil engineering at U.C. Berkeley in 1984. His thesis topic was published as an EERC report (83-24), *Design Considerations for Shear Links in Eccentrically Braced Frames*, co-authored with Egor Popov. Working for Degenkolb Engineers after graduation, he is now a Senior Principal in the firm.

notes and equations. It's just as Helmut and others have described it. You showed up on time for a Bertero lecture and you were already behind. We saw that the notes looked like they were about concrete so we figured this professor was in the right classroom and was filling in for Steve Mahin. It was the week after the Imperial County earthquake, and Professor Mahin was busy down there studying the Imperial County Services Building.

Then, getting my master's, I took the earthquake engineering class Professor Bertero taught. He started off quiet-spoken and calm. Ninety minutes later he was screaming at us, sweating, and pounding on the table because he was so engaged with his subject.

On the final exam, which required a day to do right but you only had an hour, I followed what I was sure were the right steps in using inelastic spectra. My answer was off the chart, literally it was not a result that would land anywhere on the graph. I checked my work and finally went up to him to hand in my exam and ask him where I had gone wrong. He looked at me and said, "What mean this?" [Laughter] That is one of the phrases that is a beloved "Berteroism." I said, "That's what I want to know: what mean this?" He said, "You must remember the long pulse." This was about 1981 or 1982, before the long duration pulse was a commonly accepted near-field ground motion phenomenon.

Professor Popov was my primary advisor. At that time he had a research proposal with you, Professor Bertero, to do some research on steel shapes as boundary elements in reinforced concrete shear walls, but it wasn't funded by NSF. Had it been funded, I would have done

my master's work for the two of you, and I can only imagine how hard it would have been. As a result, I worked under Popov on the eccentric braced frame.

There was a test going on in Davis Hall by a visiting professor from Africa on shallow beam-to-concrete connections. I had to be there all day helping as a lab assistant. I remember you, Professor Bertero, climbing up on top of the specimen. Meanwhile, Professor Popov motioned to one of us to bring the forklift over. He climbed on, and when Professor Bertero was just finishing his climb to get on top, Professor Popov smiled at him and casually stepped off the forklift. Those were fun days.

Reitherman: Later on, after the Northridge earthquake in 1994, you were the research manager for the SAC Steel Project,[59] when Steve Mahin was overall manager and Ron Hamburger was in charge of the guidelines development. Around the table here today, we see Professors Anderson, Krawinkler, and Whittaker, who did a lot of work on that project too. I think your exact title, Jim, was director of topical investigations, which behind your back in the CUREE (Consortium of Universities for Research in Earthquake Engineering) office we called it "tropical investigations" and joked that you were off in Tahiti enjoying yourself. You came into contact with both

---

59. SAC was a joint venture of the Structural Engineers Association of California, Applied Technology Council, and Consortium of Universities for Research in Earthquake Engineering. The SAC Steel project was primarily funded by the Federal Emergency Management Agency.

Professor Popov and Bertero in that project, somewhat as their boss rather than reporting to them as you did as a student.

Malley: Professor Bertero was teamed up with Andrew Whittaker for a series of steel tests. At that time, Andrew was a research engineer at EERC. The SAC project was not only a practical problem-solving effort, it was a reunion of everybody who had done earthquake research on steel for the past several decades. Every few months all the researchers got together, and it was quite a seminar, with Professors Bertero and Popov and former students like Helmut and other researchers from around the country. It was a great learning experience.

Reitherman: When you went to work for the Degenkolb firm, how did you apply what you learned at the university? Was it an easy transition, or a difficult one?

Malley: One summer before I graduated, Egor's funding ran out on a research project, and he said he would get me some kind of summer job. He called Henry Degenkolb and suddenly I had a job there.

Much of what we learned in class and in the laboratory in school fit right into what we were doing in practice. The Degenkolb firm looked at the recent hires, the recent students, to push the practice forward with new techniques and ideas. Much of what we learned from Professor Bertero in his seismic design class around 1980 still hasn't really come to fruition in the building codes. Many of us have used those concepts outside of the code setting for a long time now.

Andrew Whittaker

Reitherman: Andrew, tell us your tales. When did first you meet Professor Bertero?[60]

Whittaker: It was my first semester here in the United States at Berkeley. I had graduated from the University of Melbourne and worked a few years in Australia for the Connell Wagner Group, before doing my master's and PhD at Berkeley. This was 1984, and I was impressed with how meticulous the Berkeley professors were. Professor Graham Powell started precisely at ten past the hour, right as scheduled. Professor Anil Chopra started precisely at ten past the hour. Then I was sitting there for the first time in Professor Bertero's class. I recall I was sitting next to Mike Engelhardt, now a professor at the University of Texas. We had walked in a couple of minutes prior to the start time, and, to corroborate everyone else's story, the blackboard was already covered. I can add a little detail to the blackboard story. The top left-hand margin of the board had the reading assignments for the week, the bottom half of the margin had the homework assignments. That left a lot of square footage for him to pack in small, handwritten notes, equations,

60. Andrew Whittaker was a PhD student of Professor Bertero from 1984 to 1988. His thesis topic was published as EERC report 88/14, *An Experimental Study of the Behavior of Dual Steel Systems*, co-authored with Chia-Ming Uang and V. V. Bertero. After working as a structural engineer for Forell-Elsesser Engineers, Whittaker was the associate director of EERC and later PEER before becoming a professor in the Department of Civil, Structural, and Environmental Engineering at the University at Buffalo.

and diagrams. At the official class start time, ten past, this professor was calling students by name and asking them questions.

You know, Vit, I think I learned more from you that semester than from anyone in any similar period of time since. I should mention a few hiccups in the learning experience, however. One was "seldo-acceleration," a term with which we were unfamiliar. From Anil Chopra's class we had heard about pseudo-acceleration and spectral acceleration but no mention of "seldo-acceleration." There wasn't one person in the class who was brave enough to put up a hand and ask. We were afraid there would be a question about it on the exam. After about three weeks, we noticed you had written "pseudo-acceleration" on the board and only then did we figure it out.

I remember distinctly the student presentations we had to give every six or seven weeks. It struck fear in our hearts. We stayed up late the nights before to prepare our presentations.

For my qualifying exam, I had Professors Mahin, Powell, and Bertero. Steve Mahin said his question should take me about half a day, and it took me about half a day. Graham Powell gave me a question, said it would take me about half a day, and it did. Vit gave me a question and said it would take about a day or so. I spent six out of my allotted seven days on his question, working at least fifteen hours a day. When I turned in my written solutions to their questions, I had about five pages for Steve, five pages for Graham, and 120 for Vit.

The day before my oral exam, I was in Professor Bertero's office and we had a discussion, actually an argument, about various technical topics. We argued from midday to about four o'clock in the afternoon, when we were both exhausted and called it quits.

Come the oral exam the next day, I made my presentation and Graham, the chair, asked his colleagues if there were any questions. Vit put his hand up, and took up the argument of the previous day right where we had left off. After about 45 minutes, Graham broke in, said he had heard enough, and called for the next question. Far and away one of the toughest weeks in my life, but in retrospect also enjoyable. The bit of good fortune I had in my oral exam was that Professor Jim Kelly had organized a wine and cheese affair for the faculty in the department that afternoon at a quarter to four. My exam started at two. Kelly talked to Graham, I was sent outside to await their decision, and in about thirty seconds was congratulated on passing. On the way out Jim said, "It was an easy decision—either talk about you another two hours or drink wine." [Laughter] Vit was known for how long and hard he worked his candidates, but he worked them well.

Reitherman: In the early days of the American colonies, indentured servants had to work seven years to gain their freedom, and Professor Bertero has the reputation of trying to apply that same time requirement to his PhD students. How long did it take for you?

Whittaker: I think I hold the record for the shortest master's and PhD degree from Vit at four and one-quarter years.

Reitherman: There were a number of times you still had contact with Professor Bertero and worked on projects with him after you graduated with your PhD. Tell us about that period.

125

Whittaker: There were many such experiences, and all of them make me smile. When we were graduate students, we had to get our time cards signed by Professor Bertero to get paid, and he would ask us what we had done to deserve his signature on that card. In the SAC Steel Project, I was the principal investigator of the testing task Vit was working on. As usual, he worked harder than anyone. One day he came to me with a time card that needed the PI's signature. And it gave me huge satisfaction to ask with a straight face, "Tell, me Vit, exactly what have you done this week?" He was stunned, until a second later when I couldn't keep from smiling any longer.

Another story concerns the 1995 Kobe earthquake. Vit and I spent some time there along with other EERC-Berkeley investigators. The EERC team came back from Japan and put on an hour-and-a-half presentation on the campus. I scheduled Professor Bertero last, and emphasized that he had fifteen minutes, ten slides. It's difficult for a Bertero presentation to end up less than an hour and a half. At the lecture hall, I saw him walk in with two slide carousels. If you recall, there were two sizes. His were the larger, 140-slide carousels. So that was 280 slides. I introduced him on schedule when there was fifteen minutes left. Fifteen minutes later the doors opened and the next group of students had started to come in to use that lecture hall, ending Vit's presentation when he was four slides into a 280-slide presentation. We have to put into the record, Vit, the fact that you have been legendary for your two-carousel barrage of facts and images. And the text slides were photographically copied from printed-out text that had your trademark red-ink circles drawn as bullets.

Eduardo Miranda

Miranda: My first meeting with Professor Bertero wasn't in a classroom at Berkeley.[61] It was a few days after the 1985 Mexico earthquake in Mexico City. He showed up at the Institute of Engineering, with hard hat and two Nikon cameras. I have to explain that these 35 mm cameras were not the light pocket-size ones of today. They were very heavy, with lots of metal mechanical parts. And he carried a huge briefcase. Because I was doing some research on the statistics of the damage, and since my professors were all so busy because of the earthquake and I could speak English, I became a tour guide for the professors who came. It was perhaps the best job I have ever had, taking these experts to look at damaged buildings. One day it was Helmut Krawinkler, the next Steve Mahin, then Mete Sozen, and so on. I would just take them to the buildings and ask them what happened to the building and listen. When I asked Professor Bertero that question I got my first experience of his mannerism of putting his hand to his tilted head as if the thinking was almost painful, as he tried to come up with the precise answer. For three days I toured him around. I called on him at his hotel at his desired 6:30 am time—quite early by Mexican standards. In his hotel room you

---

61. After graduating from the National Autonomous University of Mexico (UNAM), Eduardo Miranda taught there before doing his master's and PhD work under Professor Bertero from 1986 to 1991. His thesis was on the "Seismic Evaluation and Upgrading of Existing Buildings." He is a professor in the Department of Civil and Environmental Engineering at Stanford University.

could see he had papers all over the room that he had already been working on.

I mentioned that he had that big briefcase. I kept asking him, "Professor, may I carry that for you?" but he declined. The elevators were out, so we were doing a lot of walking up and down stairs. He went up the stairs two at a time.

About 4 pm one day we were looking though some fifteen-story buildings—of course he insisted on going up the stairs so he could walk through every floor. I asked him again if I could carry his briefcase, and he finally said okay.

That thing was completely full of papers—it must have weighed fifteen or more pounds, and I got tired just carrying it for a little while.

The third trip he made back to Mexico City, he was giving a special lecture at the university, which was my first exposure to the double carousel slide presentation. I remember his "in other words" phrase that he uses occasionally. The room was packed with people wanting to hear him talk. It was the connection with Professor Bertero's trips to Mexico that led to my going to Berkeley for my PhD.

When I arrived at Berkeley in 1986, Professor Bertero had an NSF-funded project on the Mexico earthquake, in particular on the problem of evaluating and upgrading existing buildings. But my first task was to take some drawings of Mexican school buildings and do a nonlinear analysis of them. I had never done a nonlinear analysis, but he made it sound like I had to get it done by the following week.

I thought my task would be to analyze collapsed buildings to see why they had collapsed, but he had me study buildings that stood up to see why they didn't collapse, even if the

first analysis said they should have collapsed. I started to learn about overstrength, which was often capable of counteracting even a major vulnerability like the short column condition. That ended up being my master's thesis.

Professor Bertero has amazing intuition, meaning the ability to draw on his detailed engineering knowledge to get straight to the critical factor. I would show him results of several weeks' work and he would say, "Eduardo, there is something wrong with your analysis." I would reply, "I don't think so. I've been doing these analyses backwards and forwards to check them." I learned later you don't say that to him. He would just frown and do his own analysis, skipping the usual code-type procedures, to show what was in error.

After the 1989 Loma Prieta earthquake, I started doing some research on instrumented buildings, one of which was the Pacific Park Plaza in Emeryville, on the inland side of I-80 by the Bay Bridge. At that time, it was the tallest reinforced concrete building in northern California. It had already been studied by Berkeley people using ambient vibrations, and with shakers, I think, by Ray Clough. And it had been analytically studied by Ed Wilson, with the analytical studies matching the vibration results very nicely. It had a period of something like 1.6 seconds, verified in all three kinds of studies. Then I analyzed the strong motion records from Loma Prieta, and came up with 2.4 seconds. So here I came to my weekly meeting, afraid. Professor Bertero said, "Eduardo, how can it be?" I had learned from him that in addition to your detailed analysis, you should resort to first principles. We looked at the trace of the displacements and counted off the number of seconds per cycle and saw clearly that the period was longer than previously found. It is still somewhat of a mystery.

Reitherman: Did you begin to expect you would have a big earthquake, like 1985 Mexico and 1989 Loma Prieta, to feed you a steady diet of earthquake research material every five years or so?

Miranda: Actually, when the 1985 Mexico earthquake occurred, I was working on my undergraduate engineering thesis on wind engineering and the revision of the Mexico City building code wind provisions. I had been working on it for a year and half, then immediately after the earthquake I was assigned to the tour guide job and collecting damage statistics. After six months my advisor wanted me to shift to a seismic topic, but I was able to convince him I was very close to finishing my degree requirements with the wind engineering topic, so I had a break from the earthquake work.

When I got to Berkeley, something similar happened. I was working on retrofits related to the Mexican earthquake. I was here at EERC at the Richmond Field Station when Loma Prieta happened about five o'clock in the afternoon. It didn't feel like much of an earthquake at this site. But then immediately I was dragged away from my ongoing research. First, Professor Bertero sent Andrew Whittaker and me to Santa Cruz, which I recall was a four-hour drive because of road closures. Then we worked on the Cypress Viaduct collapse in Oakland for several days, which a lot of people were studying.

My first Berkeley course with Professor Bertero was CE 243, Comprehensive Design of Structures. To this day, I think that is the perfect title for a Professor Bertero structural engineering course. "Comprehensive" meant he could throw in everything—steel, concrete, seismic, gravity, limit design.

Once he had to miss a class and scheduled a three-hour make up lecture for the missed two-hour class. So, he was already ahead one hour. But then we noticed that after the first time, the make-up lecture became regularly scheduled. We asked each other, "Make up class? Make up for what?"

People have talked about his blackboard manner. You could hear the chalk hit the blackboard as he poked it vigorously as he wrote.

I remember the day when I got his signature on the title page of my dissertation. We called his signature the square root, because it looks like that. I was just finishing my abstract, and he kept making changes to it. It was the last day when I could file the dissertation. Even on the last day I had to make changes. When finally I got the "square root," I hustled to the campus and got it in just barely by the deadline. I was so nervous I locked myself out of the car.

Others have commented on the student presentations. At that time, there was no computer projection like the PowerPoint presentations of today. Ian Aiken and Andrew Whittaker had figured out how to do white lettering on a blue background, taking photographic slides, getting them processed in one day, and that raised the bar on presentations. We called those presentations "a time for public humiliation." No matter how good your research and your presentation, every possible flaw would be exposed. He might interrupt you and say, "Big numbers, Eduardo. You need big numbers. People in the back can't read your numbers." I remember Professor Bertero would advise us not to put too much information on one slide. [Laughter]

Whittaker: Advice from the master! [More laughter]

Miranda: Yes, his slides were packed with information, underlining, red bullets, etc. You don't quite follow your own advice, Professor Bertero, but your presentations have always been memorable.

Yousef Bozorgnia

Reitherman: Yousef Bozorgnia[62] is out of the country at the moment and couldn't be here today, but he is another of the former students Professor Bertero thought would make a good addition to this gathering. Yousef's connection with Professor Bertero falls chronologically at the end of this discussion. Yousef has provided some text material on the same topics as are being discussed in person here today.

Bozorgnia: My first contact with Professor Bertero was taking his Plastic Design of Structures course and his Seismic Resistant Design course. Both were very time consuming, but worth it.

---

62. Yousef Bozorgnia received his MS and PhD degrees from U.C. Berkeley in civil and environmental engineering, taking courses from Professor Bertero in 1978 and 1979. His PhD advisor was James Kelly and his topic was seismic isolation. Prior to his current position as Associate Director of the Pacific Earthquake Engineering Research Center, he practiced engineering with Exponent (Failure Analysis Associates) and other firms.

In my interactions with Professor Bertero, two earthquakes influenced our joint work tremendously: the 1989 Loma Prieta and the 1994 Northridge earthquakes. After the 1989 earthquake, there were many activities in seismic retrofit design of existing buildings and bridges in California. While I was in the consulting engineering profession, Professor Bertero was peer reviewer of the Hayward City Center base isolation upgrade and the retrofit of a tall building in San Francisco. We had to spend long hours to incorporate his comments into the design, but again it was worth it.

In the late 1990s Professor Bertero and I received research funding to study use of damage indices for real-time post-earthquake damage assessment. I remember that Professor Bertero and I decided to send a paper to the ASCE *Journal of Structural Engineering* on damage spectra. I carefully drafted the paper, revised it a few times, polished it, and passed him the draft to review. After a few days we met again. He said, "Your draft was very good, and I have only minor comments." I was naturally happy that my draft was "very good" and that he had only "minor comments." He passed me a copy of the paper with his markups. When I looked at it, I saw more red ink comments than the original black ink text. It took me two or three iterations, and two or three weeks, to incorporate his original comments, plus his comments on my iterations. But when it finally met with his satisfaction, it went through the review process and was very easily published. I have noticed that he doesn't spare himself from his critical editing either.

That work led to a discussion I had with Professor Bertero in 2000 on the idea of a book that would bring in some of the more recent thinking in the earthquake engineering field. That resulted in the anthology we produced in 2004.[63] The title expresses a historical viewpoint. The "engineering seismology" in the title refers to the early term for what was to become the earthquake engineering field. The "performance-based engineering" phrase indicates the current work being done. He and I wrote the first chapter, "The Early Years of Earthquake Engineering and Its Modern Goal," which also indicates the historical background to the book.

Convincing the potential authors to write various chapters was a relatively easy task, as soon as they found out that Professor Bertero was an editor of the book. One difficult task was deciding on the title. The original title we proposed to the publisher was *Recent Advances in Earthquake Engineering*. The publisher wanted to put "Handbook" into the title, but Professor Bertero didn't think the book fit the definition of that word, and he refused.

I recall a time during the extensive editorial process to put the book together and have everything edited properly when Professor Bertero gave me a phone call at home on a Saturday. "Yousef, I am very concerned about an equation in a particular chapter. Let's meet." I met him at the EERC library. He was sure there must be a typo of some kind in the equation, even though it had already passed the review of the author of that chapter. We

---

63. Yousef Bozorgnia and Vitelmo V. Bertero, editors, *Earthquake Engineering: From Engineering Seismology to Performance-Based Engineering*. Boca Raton, LA, CRC Press, 2004.

derived the equation independently, and, as usual, Professor Bertero was right.

Bertero's Closing Comments

Reitherman: Your students have turned out well, Professor Bertero. You must be pleased with the results, in other words, how they have made something of themselves in their careers.

Bertero: Yes, I would like to say a few words. I learned when I was a student that I learned the most from the professors who were willing to really listen and help. Even if the professor demands hard work, this makes the best situation for the student. To see now you successful people around the table, in your careers—in the university or in practice—makes me happy.

Reitherman: Your former students all agree that you were a very tough professor. Any rebuttal to their good-natured stories about that?

Bertero: Yes, I was tough on them, but not because I liked to see them suffer. A good teacher is demanding. Yes, I know I have been demanding, but I hope they appreciate what I was trying to do for them.

Now that I am retired for several years, I see the history of the whole field more clearly. And now my students are retiring. If there is one thing that worries me, it is that people are not looking at the history. They are not learning from what was done in the past. They are repeating errors, which is a waste of time.

Krawinkler: There is a part of you in all of us, in our teaching and our research. We have been educated by you and it is reflected in the way we work, and we are passing it on to the next generation. You don't know all of our many students, but they know you.

Mahin: This is like genealogy. I look at my students and sometimes think of them as "Bertero's great grandchildren." It carries forward.

Bertero: I appreciate all you say here today. I am glad to see you all be leaders now. I have confidence for that reason in the generation you have taught, the ones who will carry on.

Photographs

Professor Bertero grew up in this house in the city of Esperanza (built in 1928). Since the deaths of his parents, the building has been used as a school for teaching English. The photo pictures Bertero at the site in September 2006, while he was visiting the city.

A portrait of Vitelmo Bertero's parents, Victorio Bertero and Lucía Gertrudis Risso Bertero, as a young married couple.

Portrait of Vitelmo Bertero, left, at age two, and his four-year-old brother, Humberto.

Brothers Humberto and Vitelmo Bertero and their families. Back row, left to right, with relationships to Professor Bertero: Humberto Bertero (brother), Vitelmo Bertero, Nydia Barceló Bertero (wife), Victorio Bertero (father). Front row: Sara Bertero (niece), Nelly Bottai de Bertero (sister-in-law), María Teresa Bertero (daughter), Marta Bertero (niece), Eduardo Bertero (son), Lucía Gertrudis Risso Bertero (mother), Tito Bertero (nephew). (1953)

The main entrance of the Colegio San José, *where Bertero attended elementary and secondary school.*

The boarding house in Rosario, Argentina, where Bertero lodged as a student while attending the Universidad Nacional del Litoral.

The Faculty of Mathematical, Physical-Chemical, and Natural Sciences Applied to Industry (Facultad de Ciencias Matemáticas, Físico-Químicas y Naturales Aplicadas a la Industria) *of the* Universidad Nacional del Litoral *in Rosario, Argentina, from which Bertero graduated with a civil engineering degree. (1947)*

A group photo of U.C. Berkeley faculty members at the 1997 EERC-CUREE Symposium in Honor of Vitelmo Bertero.

Left to right, front row: Joseph Penzien, Egor Popov, Vitelmo Bertero, Boris Bresler, Hugh McNiven, Ray Clough. Back row: Gregory Fenves, James Kelly, Armen der Kiureghian, Alexander Scordelis, Filip Filippou, Anil Chopra, behind Chopra is Nikos Makris, next to him are Karl Pister, Edward Wilson, Stephen Mahin, Jack Bouwkamp, Jack Moehle, Andrew Whittaker.

In September 2006, Professor Bertero was an honored guest at the festivities commemorating the 150th anniversary of the founding of the city of Esperanza. Here he meets with the mayor, Rafael A. DePace. Bertero notes that "the mayor and his six deputies started to ask me questions about my activities in the USA, and particularly about earthquake engineering. After two hours of discussion, the mayor informed me that I had been declared a Distinguished Citizen (Ciudadano Ilustre) of the city and was a featured guest at the main public ceremony the next day, which included a parade."

Nydia Ana Barceló Vilas and Vitelmo Victorio Bertero were married in 1949.

Vitelmo and Nydia Bertero and their children on the occasion of their fiftieth wedding anniversary. Standing around them, left to right, are Edward, Richard, Mary Rita, Robert, María Teresa, and Adolph. (1999)

Nydia and Vitelmo Bertero in 1999 on their fiftieth wedding anniversary.

Professor Bertero at the Engineering News-Record and Applied Technology Council award ceremony in April 2006. He was honored as one of the top thirteen U.S. earthquake engineers of the twentieth century. Professor Bertero was previously honored by Engineering News-Record in 1990 as Man of the Year.

A

Aeronautical engineering, 36

Aftershocks, 103–104

Agbabian, Mihran (Mike), 49

Aktan, A. E., 60

Aktan, H. M., 60

Almant, R., 52

Alonso, L. J., 52, 92

American Concrete Institute (ACI), 43, 85, 91, 94, 110

American Institute of Architects (AIA), 62

American Institute of Steel Construction (AISC), 117

American Iron and Steel Institute (AISI), 117

American Society of Civil Engineers (ASCE), 43, 79, 110, 129

Anagnos, Thalia, xi

Anchorage, Alaska earthquake (1964), 41, 64, 78, 92, 114

Anderson, James C., 52, 60, 108, 113–115, 118

Antebi, Joseph, 37

Aoyama, Hiroyuki, 57

Applied Technology Council (ATC), 70–71, 85, 139

Architecture, educational curriculum, 8, 22, 60–62, 73, 91

Argentina, earthquakes in
Caucete (1977), 18, 74, 77, 99
San Juan (1944), 17–19

Arnold, Christopher, 62, 73, 93

Aroni, S., 52

Atalay, M. B., 52

ATC 3-06, 70–71, 85

Awards. *See* Honors and awards

B

Baker, A. L. L., 42

Baker, J. F., 42, 110

Baluchistan, India earthquakes (1931, 1935), 18

Bancroft Library, Regional Oral History Office, viii

Barceló, Adolfo (father-in-law), 12

Baron, Frank, 75

Base isolation, 78, 129

Baum, Willa, viii

Beam-column connections, 55–56, 78, 95, 99, 116, 121, 123

Beedle, Lynn, 48

Benjamin, Jack R., 26, 33, 72, 117

Benjamin, Jack R. Associates, 27, 33

Benuska, K. L. (Lee), 72, 78

Bertero, Adolph (son), 56, 138

Bertero, Domingo, 23

Bertero, Edward T. (son), 23, 26, 56, 109, 111, 134, 138

Bertero, Humberto (brother), 1, 23, 134

Bertero, Lorenzo (paternal grandfather), 1

Bertero, Lucía Gertrudis Risso (mother), 1, 134

Bertero, María Teresa (daughter), 23, 26, 56, 109, 134, 138

Bertero, Marta (niece), 134

Bertero, Mary Rita (daughter), 26, 56, 109, 138

Bertero, Nelly Bottai de (sister-in-law), 134

Bertero, Nydia Ana Barceló Vilas (wife), 12, 22, 109, 134, 138, 139

Bertero, Raúl, 23, 52, 60

Bertero, Richard (son), 56, 110, 138

Bertero, Robert (son), 26, 56, 109, 138